SOIL FERTILITY, FERTILIZERS AND AGROCHEMICALS

The Authors

Dr. Praveen Kumar Jaga obtained M.Sc. (Ag) in Soil Science and Agricultural Chemistry from JNKVV, Jabalpur in 1999 and Ph.D. in Soil Science and Agricultural Chemistry from Dr. Bhimrao Ambedkar University, Agra (U.P.) in 2003. He is presently rendering his services in the Department of Soil Science and Agricultural Chemistry, College of Agriculture Ganj Basoda, District, Vidisha (M.P.) a constituent college of JNKVV, Jabalpur as an Assistant Professor. He served as Senior Research Fellow and Senior Research Associate at RBS, College Bichpuri (Agra) and Indian Institute of Soil Science (IISS), Bhopal. He is recipient of Bharat Shiksha Ratan Award-2013 under by GSHE, New Delhi, Young Agriculture Scientist Award-2008 Annadata Channel Foundation Award-2007, Bhumi-Nirman Awards-2006 and "Certificate of Excellence" for farmers through Annadata Channel on ETV-MP/CG. He has 15 years experience in research, teaching and extension and has more than 244 publications and three books of national and international level.

Dr. Bharat Singh graduated from R. B. S. College, Bichupri, Agra with first class. He obtained M. Sc. (Ag.) from Raja Balwant Singh College, Bichpuri, Agra with first class. He secured Ph.D in Soil Science & Agricultural Chemistry from B. R. Ambedkar University, Agra (UP). He has worked 10 years in M.P. Science and Technology, Bhopal as Scientific Officer. He has an experience for over 15 years in teaching research and extension. He guided seven research students majoring in Soil Microbiology and published about 45 scientific articles in national and international Journals.

Cataloging in Publication Data—DK
Courtesy: D.K. Agencies (P) Ltd. <docinfo@dkagencies.com>

Jaga, Praveen Kumar, author.
A handbook of soil fertility, fertilizers and agrochemicals / authors, Praveen Kumar Jaga, Dr. Bharat Singh.
pages cm
Includes bibliographical references.

ISBN 978-93-5130-955-0 (International Edition)

1. Soil fertility. 2. Fertilizers. 3. Agricultural chemicals. I. Simòha, Bharata, author. II. Title. III. Title: Soil fertility, fertlizers and agrochemicals.

S596.7.J34 2016 DDC 631.422 23

Published by : **Daya Publishing House®**
A Division of
Astral International Pvt. Ltd.
– ISO 9001:2008 Certified Company –
4760-61/23, Ansari Road, Darya Ganj
New Delhi-110 002
Ph. 011-43549197, 23278134
E-mail: info@astralint.com
Website: www.astralint.com

Laser Typesetting : **Classic Computer Services**, Delhi - 110 035

Printed at : **Replika Press Pvt. Ltd.**

SOIL FERTILITY, FERTILIZERS AND AGROCHEMICALS

Dr. Praveen Kumar Jaga
Dr. Bharat Singh

2016
Daya Publishing House®
A Division of
Astral International (P) Ltd
New Delhi 110 002

Preface

One of the major problems of Indian agriculture is low crop productivity. The appropriate technology of crop production and effective use of input like seed, fertilizers, irrigation herbicides *etc.* is the need of the day for any developing nation. This book is written in the format to fulfill the requirement of textbook practical manual and preparation for competitive examination.

Soil, plant, and water are the major ingredients of agriculture system. Analysis of soil, plant, and water are utmost important. This manuscript is a compilation of tested and widely used analytical methods of characterizing soil, plant and water. Brief explanations of the important steps are also incorporated.

We hope the technical information will be useful to students, teachers, scientists and analyst studying/working in the field of agriculture specially soil science, agronomy, microbiology, and forestry. Efforts are made to make the manuscript self explanatory and useful, however, any suggestions for improvement will be highly appreciation. The authors are highly pleased and expressed his gratitude for encouragement.

The authors are grateful to Dean, College of Agriculture, Ganj Basoda, Vidisha (MP) for keen suggestions for improvement of this book.

The authors will highly grateful to Dr. V.S Tomar, Vice-Chancellor, JNKVV, Jabalpur, Dr. S.S. Tomar, Director, Research Services, JNKVV, Jabalpur for suggestions and constructive thoughts for revision and improvement of the book. The Authors acknowledge the every source of information from where he has drawn technical matter.

Praveen Kumar Jaga
Dr. Bharat Singh

Contents

1
Soil Fertility

Literally, the world 'fertile' means bearing abundantly and a fertile soil is considered to be one that produces abundant crops under suitable environment conditions. A distinction should be made however, between soil productivity and soil fertility. Soil productivity is basically an economic concept and signifies the capacity of soil to produce specified plant or plant part or a sequence of plants under well defined and specified systems of management inputs and environmental conditions, *viz.* climatic condition. It is measured in terms of out puts or harvests in relation to production factor for a specific kind of soil under a physically defined system of management. Soil fertility refers to the inherent capacity of a soil to supply essential nutrient elements to the plants in adequate amount and in right proportion for their optimum growth. It is one of the key components to determine productivity. It is defined as the quality of a soil that enables it to provide essential chemical elements in adequate quantities and proportions for the growth of specified plant. The essential nutrient elements comprise the key components of soil fertility vis-à-vis plant nutrition and therefore these are referred to as plant nutrients or nutrient elements.

Types of soil fertility

- **Inherent or natural fertility:** The soil, as a nature of them, contains more some nutrients which are known as inherent fertility. Among the plant nutrients Nitrogen, phosphorus and potassium is essential for the normal growth and yield of crop. Indian soil contains 0.3 to 0.2 % N, 0.03 to 0.3 % P and 0.4 to 0.5 % K. The inherent fertility has a limiting factor from which the fertility is not decreased.
- **Acquired fertility:** The fertility developed by application of manures and fertilizers, tillage, irrigation etc. is known as acquired fertility. The acquire fertility has also a limiting factor. It is found by experiment that the yield does not increase remarkable by application of additional quantity of fertilizers. So it is necessary to apply fertilizer on the basis of nutrient content of a soil and it is estimated by soil testing.

Causes of decreasing the fertility: Soil is the store house of plant nutrients. Losses of plant nutrients from the soil are the main causes of decreasing the fertility of soil. Plant nutrient are lost from the soil by the following ways.

(i) **Removal of plant nutrients by harvested crop**: Plant absorbs nutrients from the soil and stores them in their different parts. The crops remove large quantity of N and K and relatively small quantity of P. Four to five percent of total N is lost from soil per acre annually through the harvested crops. The loss can be reduced by adding farm waste materials to the soil.

(ii) **Removal of plant nutrients by weeds:** Weeds compete with crops for mineral nutrients. Competition begins when the supply of plant nutrients falls below the requirements of both weeds and crops. Weeds by nature of them grow fast and remove the plant nutrients from the soil,

(iii) **Losses of nutrients by soil erosion***:* Erosion is the physical removal of top soil by water and wind. Plant nutrients, particularly N remains on the upper layer of soil. When erosion is severe, the nutrient is lost along with soil and the fertility of soil decrease accordingly.

(iv) **Losses of nutrients by soil erosion:**Fertilizers both straight and mixed are soluble in water and as such hay are liable to loss by leaching in rain water or irrigation water. Leaching loss is more acute in sandy soil and bare soils. N is mainly lost from soil by leaching.

(v) **Losses of nutrients by gaseous form:** N is generally subjected to loss in gaseous form. Losses of N occur due to denitrification.

Factor affecting soil fertility

The factor affecting soil fertility may be of two types (A) Natural factor (B) Artificial factor. The natural factors are related to the proper use of land. The factors affecting the fertility of soil are as follows.

Parent materials: The property of soil depends on the property of parent rock. If the parent rock contains more nutrients, the soil developed from rock contains more nutrients. The soil developed from calcareous rock contains more P than the soil which is developed from granite rock.

Climate and vegetation: Rainfall and temperature has an effect on soil fertility. In heavy rainfall areas, the nutrients are lost by leaching. As result of which the fertility of that soil becomes low. Besides these, the upper layer is eroded which decrease the soil fertility.

Topography: The fertility of soil is also dependent on the topography of soil. Leaching and erosion is most common in sloppy land. As a result of which, the fertility of that soil becomes low.

Inherent capacity of soil to supply plant nutrient: The nutrient contents of a soil vary according to the nature of soil. The soil which contents much quantity of nutrient becomes more fertility.

Physical condition of soil: Aeration and movement of water is good in the soil containing adequate amount of organic matter and this type of suitable conditions of soil is beneficial for the growth of plants. The physical condition of soil should be suitable for the growth and development of plants. It is essential for proper supply of oxygen in the soil. Improper supply of oxygen is unsuitable for the growth of plants as well as for the proper function of soil organism. Suitable physical condition of the soil increases the water holding capacity of soil which is favourable for the growth of plant. The fertility of soil depends mostly on the texture and structure of soil.

Soil age: The soil developed earlier losses its fertility gradually because the fertility of the soil decreases by the process of leaching and weathering in course of time. Besides this, cultivation of crops without manuring decreases the fertility of soil.

Micro-organism and soil fertility: various types of organism live in the soil. The soil organism brings the unavailable nutrients into the available form.

Availability of plant nutrients: The nutrient of the soil must be in the available form of plants. The plant do not absorbs nutrient if it is not suitable in water. Super phosphate applied in acid soil is converted into iron or aluminium phosphate which is not soluble in water. As a result, P remains in the soil in unavailable form to plant.

Soil composition and fertility: The plant absorbs the nutrient from the soil. The nutrient of minerals becomes available by weathering. The soil containing more organic matter becomes more fertile. The sandy soil is less fertile, where as loamy soil is more fertile.

Organic matter and soil fertility: The fertility of soil increases if the soil contains more organic matter. Organic matter contains plant nutrients. Besides this, Organic matter improves the physical condition of soil. Decomposition of Organic matter increases the N content of soil.

Soil erosion: Erosion is the physical removal of top soil by water and wind. As such it decreases the fertility of soil because the nutrients remaining in upper layer of soil is lost by erosion and the fertility of soil decreases accordingly.

Cropping system: Cultivation of same crop year after year in the same field decreases the fertility of soil. Crop rotation increases the fertility of soil.

Favourable environment for root growth: The suitable condition for growth of plant depends on physical, chemical and biological condition of soil. Soil contains 25 % water and 25 % air by its volume and this condition is favourable for good aeration. The bad aeration in the soil is not good for the growth of the soil.

Management of soil fertility

Efficient nutrients management is the prime concern in the management of optimum soil fertility. This has received the maximum attention n during the last four decades as is evident from the fertilizer consumption pattern and its contribution in increasing crop production. During the early stages, the increase in food grain production was mainly due to increase in area of cultivation but after 1970-71, it has been largely due to increase in the productivity. In a successful management of soil fertility following aspects need consideration:

- Optimum fertilizer rates
- Balanced use of fertilizer nutrients
- Efficient use of fertilizer
- Nutrient needs of cropping system and
- Integrated nutrient management

Optimum Fertilizer Rates

The amount of fertilizer needed by a soil depends upon its nutrient supplying capacity to crop and the requirement of crop. N is the most limiting factor for crop production in Indian soils because of their poor organic carbon content. The response of crops to N is universal in almost all soils of India, except in hilly region soils containing high organic matter. The chemistry of applied fertilizers is different for upland crops and submerged rice culture. The low efficiency of N utilization in rice is a matter of concern because rice alone accounts for about 40% of the total fertilizer N consumption in India. Rice, wheat and maize generally respond significantly up to 120kg N /ha through at some places responses up to 150-180 kg N /ha have also been reported (Narang and Bhandari, 1992). A leguminous crop generally requires a starter dose of about 10 kg N /ha as a bulk of the requirement is met through fixation of atmospheric N.

Deficiency of P is also wide spread in Indian soils. Soil testing low and medium in P require the application of P for optimum yields. Wheat crop is highly responsive to P application. In general, it respond S to P applied up to 60 kg P_2O_5 /ha while in mixed black and red and yellow soils, its response to P continue up to 120 kg P_2O_5 /ha. Similarly, rice also responds generally up to 60 kg P_2O_5 /ha. However, in medium black soil very low in Available P it responds up to 90 kg P_2O_5 /ha. Response to K application on crops is modified by the amount of reserve K and several other factors. Crop response to K is large on laterites, red and yellow and mixed red and black soils. Crops grown on coarse textured alluvial soil and some shallow black soils need K fertilization for optimum crop production.(Subba Rao and Srinivasa Rao,1996). Sulphur deficiency is also becoming wide spreading in different soils of India. Differential response to sulphur application of different crops has been reported. In Punjab and erstwhile UP, 20 kg S / ha to groundnut, linseed, pigeon pea and sorghum, 30 kg S /ha to ground nut, wheat chickpea, black gram and maize and 40 kg S /ha to green gram and lentil corrected the S deficiency in crops. In black soils, 30 kg S /ha to green gram, soybean and 50 kg S /ha to potato and 20 kg S /ha to ground nut efficiently corrected S deficiency. Rice and maize responded significantly up to 60 kg S /ha whereas mustard, wheat and berseem responded up to 90 kg S /ha. The beneficial effects of S application ranging from 2 to 12 q/ ha in different cropping systems have been reported (Tandon, 1989). All acids soils and many sodic soils are inherently deficient in Ca and /or Mg and crops respond to the application of these elements. Among the micro-nutrients disorders of crops, zinc deficiency is critical in many states of India. Responses to Zn differ widely among crops in each state as well as across the states. *Kharif* crops generally give a better response to Zn application than *rabi* crops; the magnitude is influenced by several factors such as crops and their cultivars, soil types deficiency status of Zn and other nutrients, soil environment, climate and cultural practices, etc. The deficiency of Zn, Cu, Fe and Mn are common in soils having high pH, high Ca CO_3 contents. On the other hand, deficiency of B and Mo are more prevalent in acidic soils; B is deficient in highly calcareous soil also.

Balanced Use of Fertilizer Nutrients

Nutrients are removed continuously by the crops from the soils and their replenishment through fertilizers and manures is essential. The ideal ratio of nutrients for optimum growth depends on the crop and soil. When the added fertilizer contains nutrients in less or higher amount than that required by the crop, the imbalance thus, caused results in poor crop growth. The application of P along with N increases the nitrogen use efficiency. Some times even with NP application, decrease in yield occurs because K becomes the limiting factor. Similarly, with optimum dose of NPK, the deficiency of Zn may appear. Soil test based fertilizer recommendations ensure balanced use of fertilizers and increase yields and profits. Balanced fertilizers encompass, beside major nutrients, secondary and micro-nutrients whose deficiency appear in different soils in various agro-eco-regions due to intensive cultivation.

Efficient Use of Fertilizer

Increasing nutrient efficiency is the key to the management of soil fertility. The proportion of the added fertilizer actually used by plants is a measure of fertilizer efficiency. Soil characteristics, crop characteristics and fertilizer management techniques are the major factor that determines fertilizer efficiency.

Soil characteristics

Nutrients status of soil: The response of any crop or a cropping system to added nutrient depends largely upon the inherent capacity of soil to supply that nutrient as per the requirement of crop. Chemical tests have long been used to estimate the nutrient availability in soils to predict the probability of obtaining a profitable response to applied nutrients. On the basis of soil testing, soils are rated, low medium and high in plant nutrient and suitable fertilizer amounts are recommended. In a low nutrient soil, the crop responds remarkably to its application. On the other hand, in a high nutrient soil, the crops may show little or no response. In medium test soil, the response is intermediate. Soil testing helps in adjusting the amount of fertilizer and thus, improves the efficiency of fertilizer use. By demarcating the areas responding differently to different plant nutrients right type and proper amount of fertilizer can be applied to them.

Forms of nutrients: Not all pools of nutrients are available equally to crops. Some nutrient pools even do not show in soil tests but may supply considerable amounts of the nutrient during the crop growth period. For example, sandy soils high in mica may supply sufficient K to crops even though the soil tests indicate low in available K. Responses to K on such soils are not expected in a short period but may show up in long term experiments.

Nutrient losses and transformations: The amounts of nutrients estimated by soil tests may not be entirely available to plants because of their leaching, volatilization, denitrification and transformations to unavailable forms. Leaching losses are important for nitrate nitrogen because it is not held by exchange sites in the soil, it is lost. Such losses are of particular significance in sandy soils and in situations if heavy rain or irrigation follows its application. In acid soil, leaching losses of Ca, sulphate, potassium and magnesium are more common. Volatilization of ammonia in high pH surface soils is considerable when urea is applied at the surface. Denitrification loss of N mainly occurs under waterlogged conditions prevailing during rice cultivation, particularly under higher temperatures and in the presence of easily decomposable organic materials. The conversion of a portion of available nutrients into insoluble mineral form is also important. Thus, the efficiency of added P is 20-30%. Microbial immobilization also converts temporally the soluble forms of nutrients into unavailable forms. Similarly, the efficiency of Zinc applied to soil is less than 3%. Soil characteristics play an important role in the transformation of nutrients. Soil reaction (pH) is one of the important soil properties that affect plant growth. The harmful effects of soil acidity are more due to secondary effects of high acidity or low pH in a soil are the inadequate supply of available Ca, P and Mo on one hand and the excess of soluble Al, Mn and Fe on the other. Like wise, in saline–alkali soil, the deficiency of Ca, Mg, P, Zn, Fe and Mn is very common. The fertilizer practices are therefore, to be modified according for soils with different soil reactions. The main aim of liming of acid soils and addition of gypsum to alkali soils is to change the soil pH suitable for the availability of most plant nutrients.

Soil organic matter: Soil organic matter content is generally considered as the index of soil fertility and sustainability of agricultural system. It improves the physical and biological properties of soil, protects soil surface from erosion and provides a reservoir of plant nutrients. In tropics the maintenance of soil organic matter is very difficult because of its rapid decomposition under high temperatures. The cultivation of soils generally decreases its organic carbon content because of its increased rate of decomposition by the current agriculture practices. In cultivated soils, prevalent cropping system and associated cultural practices influence the level at which organic matter would stabilize in a particular agro-eco-system. LTFEs have shown that the integrated use of organic manures and chemical fertilizers can maintain high productivity and sustainable crop production. Recent studies have indicated that periodic addition of large quantity of crop residue to the soil maintains the N and oraganic matter at adequate level even without using legumes in the rotation. The application of FYM, compost and cereal residues effectively maintains the soil organic matter (Som) due to incorporation of rice or wheat straw in to the soil instead of removing or burning it. Yields are, however, low in residue incorporated treatments due to wide C: N ratio of the residues. This ill effect, however, can be avoidable at least 20 days before seeding wheat.

Soil moisture: Fertilizer application facilitates root extension into deeper layers and leads to greater root proliferation in the root zone. Irrigated wheat fertilized with N used 20-38 mm more water than the unfertilized crop on loamy sand and sandy loam soils and increased dry matter production (Gajri *et al.*, 1989). Soil moisture also affects root growth and plant nutrient absorption. The nutrient absorption is affected directly by soil moisture and indirectly by the effect of water on metabolic activities of plant, soil aeration and concentration of soil solution. If soil moisture becomes a limiting factor during critical stage of crop growth, fertilizer application may adversely affect the yield. Fertilizer–use efficiency also depends on a good moisture supply.

Physical condition of soil-Despite adequate nutrient supply, unfavourable physical conditions resulting from a combination of the size, shape, arrangement and mineral composition of the soil particles may lead to poor crop growth and activity of microorganism. Soil N generally increase as the texture become finer. The basic requirements for crop growth in terms of physical conditions of soil are adequate soil moisture and aeration, optimum soil temperature and freedom from mechanical stress. Tillage, mulching, irrigation, incorporation of organic matter and other amendments like liming of acid soils and addition of gypsum to sodic soils are the major field management techniques that aim at creating soil physical environment suitable for crop growth. Tillage affects water use by crops not only through its effect on root growth but also affects the hydrological properties of soils. Mulching with residues, plastic film etc. influencing evaporation losses from soil by modifying the hydro-thermal regime of the soil and affects root growth and rooting pattern. Use of organic mulch also decreases maximum soil temperature in summer and minimum temperature in winter and helps in the conservation of soil moisture.

Crop characteristics

Nutrient uptake-The total amount of nutrients removed by a crop may not serve as an accurate guide for fertilizer recommendations; it does indicate the differences in their requirements among crops and the rate at which the nutrient reserves in the soil are being depleted. The nutrient uptake may very depending upon the crops and its cultivars, nutrients level in the soil, soil type, soil and climatic conditions, plant population and management practices. It is estimated that 8 tones of rice grain remove 160 kg N, 38 kgP, 224 kg K, 24 kg S and 320 g Zn as compared to a removal of 125 kg N, 20 kg P, 125kg K, 23 kg S and 280 g Zn by 5 tones of wheat from one hectare field (Tandon and Narayan, 1990).

Root characteristics-Roots are the principal organs of nutrient absorption. A proper understanding of their characteristics helps in developing efficient fertilizer practices. The absorption of nutrients depends upon the distribution of roots in soil. The shallower the system, the more dependent the plant is on fertilizers. It is well known that some plants are better scavenger of certain nutrients than others. This is mainly because of the preferential absorption of these nutrients by the roots of those plants. For example legumes have a marked preference for divalent cations like Ca^+ whereas grasses feed better on monovalent cations like K^+. The efficiency of the applied fertilizer can be improved considerably if the rooting habits of various plants during early growth stages are known. This is particularly true for relatively immobile nutrients and for situations where the fixation of applied nutrients is very high. If a plant produces tap roots system, fertilizer can best be placed directly below the seed. On the other hand, if lateral roots are formed early, side placement of fertilizer would be helpful.

Crop rotation-The nature of cropping sequence has profound effect on the fertilizer requirement and its efficiency. The crops requiring high levels of fertilizers such as maize potato may not use the applied fertilizer fully and some amount of the nutrient may be left in the soil which can be utilized by the succeeding crop. Phosphorus among the major nutrients, is worthy of consideration because only less than 20% of the applied phosphatic fertilizer is utilized by the first crop. Similarly less than 3 % of the applied Zn is used by the first crop. Crops have a tendency of luxury consumption of N and K and may not leave any residual effects unless doses in excess of the crop requirement are applied. On the other hand if sub optimal doses of fertilizers are applied to a crop, they may leave the soil in a much exhausted condition and the fertilizer requirement of the succeeding crop may increase. The legumes leave N rich root residues in the soil for the succeeding crop and thus, reduce its N requirement.

Fertilizer characteristics and management techniques-The efficiency of a fertilizer varies to a large extent on the type of fertilizer, time of fertilizer application and method of application.

Type of fertilizer- fertilizers differ in both their content and form. In the case of nitrogenous fertilizer, the nutrient may be present in ammonium, nitrate or amide form. Similarly, in case of phosphatic fertilizers, Phosphorus may be present in water soluble, citrate soluble or citrate insoluble form. The nutrient content of a fertilizer mat also differ largely. Thus, the total N content of CAN is 25 % as compared to 46.5 of urea 20.5 % of As and 82 % of anhydrous ammonia. Nitrogen may also be present as one of the components of the combined fertilizer sources such as 18 % in DAP and 13 % in $KNO_{3.}$ Similarly, SSP will supply only 16 % P_2O_5 as compared to 46 % P_2O_5 supplied by DAP and triple super phosphate. The K_2O content of KCl is 60 %. The fertilizers may also differ in their water solubility and their ability to get fixed in the soil. Nitrates fertilizers have high water solubility and are subject to leaching in high rainfall areas or due to heavy irrigation, as these do not interact with the soil constituents on their addition to soil and thus, move to greater depths. Generally, all sources of fertilizer N have been shown to be equally efficient in upland soils, but ammoniacal and amide forms of N are more efficient as compared to nitrate source of N for submerged rice soils. Phosphatic fertilizers containing phosphorus in water soluble form like DAP, SSP have been found superior for most of the crops in natural or alkaline soils as compared to citrate soluble or citrate insoluble form. Both SSP and DAP proved equally efficiently on the basis of equivalent Phosphate content but SSP has the advantage in soils low in available S and particularly for S loving crops like oilseed and pulses. Rock phosphate which has phosphate in water insoluble form has proved useful in acid soils, rice or for long duration legumes. Nitro phosphate with more than 60% water soluble P have shown promises as efficiently sources of P in alkaline soils. Most of the fertilizer K used in Indian agriculture is imported largely in the form of MOP and only about 1% as SOP. In highly leached acid soils where sulphate deficiency is expected, potassium sulphate should be preferred. For certain crops like tobacco which is chloride sensitive, potassium sulphate should be applied. Straight micro-nutrients carriers like zinc sulphate, ferrous sulphate manganese sulphate and copper sulphate have been found to be superior and economical as compared to other sources of micro-nutrients.

Time of application-Time of fertilizer application is important particularly for those nitrogenous fertilizers which tend to leach with irrigation water or rainfall. Split application of nitrogen is the most common and widely accepted practice for almost all crops. The basic concept is to apply N in two or more splits to coincide with peak period of N requirements of the crop. N application in splits doses is particularly beneficial on light textured soils in increasing the efficiency of applied N. Its application in three split doses at transplanting, tillering and panicle initiation in rice at planting, knee high stage and tasselling stage in case of maize, is usually recommended. In general, crops grown in rainy season should receive N fertilizers in split doses so that the leaching of the nutrient during heavy rains may be avoided and an adequate supply of the nutrient at critical stages is ensured For crops, like wheat, barley, *brassica* sp., which are shown in winter when the rains are scanty, best result are obtained when the entire doses of N is applied as a basal doses in medium to heavy textured soils and on light soils in two equal splits at sowing and at first irrigation. Recant studies have shown that N applied in the form of urea at pre-sowing irrigation even to a loamy sand soil gives higher grain yield of wheat as compared to its application at preparatory tillage and increases both nitrogen-use efficiency (7-34%) and water use efficiency (2-24%) at different nitrogen and irrigation levels (Sidhu *et al.*,1994).

In the case of phosphatic and potassic fertilizers as well as for zinc and copper among micro-nutrient cations, all the quantity applied at sowing gives the best results with most of the crops. However, in the case of iron and manganese where foliar sprays of these nutrients are recommended, initiation of foliar sprays is very important. In the case of manganese, foliar sprays initiated 2-3 days before the first irrigation to wheat have been found to be significantly superior to their application after the first irrigation (Takkar *et al.*, 1986). Similarly, foliar application of iron should be initiated as soon as the symptoms of its deficiency appear on crops.

Method of application-Various techniques have been developed to improve the efficiency of applied nutrients. Since, nitrates are easily leached and lost by denitrification, retardation on nitrification of ammonium containing or ammonium producing fertilizers by using nitrification inhibitors in rice fields subjected to intermitted flooding helped in increasing the nitrogen-use efficiency. Chemical nitrification inhibitors as well as indigenous materials such as Neem (*Azadirachta indica*) and karanj (*Pongamia glabra*) cakes have been successfully used except on highly coarse textured soils of Punjab. Soil application of ferrous sulphate in coarse textured soil for rice is less efficient than foliar sprays. However, for upland crops, the soil applied iron has been found to be as effective as foliar sprays.

Nutrient Needs of Cropping System

The cropping system needs should take into account the long term implications of nutrients management where direct, residual and cumulative effects of applied fertilizers are taken into consideration. The cropping pattern in irrigated ecosystem at present is a cereal-cereal system due to its higher yields, low risks and adequate support price from Government. The cropping system approach helps in optimizing the scheduling application, especially for immobile nutrients like P and Zn. The P should preferable be applied to rabi crops in the cropping sequence and has been confirmed in maize –wheat, ground-wheat and pigeon pea rotations. However, soils which are highly deficient in available P may be the exceptions where fertilizer-P application is needed to all the crops in a system.

Integrated Nutrient Management

Integrated nutrient management envisages the use of chemical fertilizer in conjunction with organic manures, legumes in cropping system, biofertilizers and other locally available nutrient sources for sustainable soil health and productivity. The combined application of organic manures and chemical fertilizers generally produces higher crop yields than when each is applied alone. This increase in crop productivity may be due to the combined effect of nutrient supply, synergism and improvement in soil physical and biological properties. FYM constituents an important component of INM for maintaining soil fertility and yield stability. The green manures are a valuable potential source of N and organic matter. *Susbania* sunnhemp and cowpea incorporated before transplanting rice partially meet the nutrients requirement of the crops. In uplands crops, the most efficient time of incorporation of green manure crops has been found to be 15 days before seeding. Besides being a source of P, K, S and micro-nutrients, green manure influences availability of the nutrients through it favourable impact on oxidation-reduction regimes, pH and increased cheletion capacity. In permeable coarse-texture soils, green manuring helps correcting the iron chlorosis in rice. The green manuring plays an important role in the reclamation of salt- affected soils and improvement in physical and biological properties of soils.

Evaluation of soil fertility

Evaluation of soil fertility consists of estimating the nutrient-supplying power of a soil. A proper evaluation of the fertility of a soil before planting a crop helps in adopting appropriate measures to make up for the short comings and ensuring a good crop production. These methods for evaluating the soil fertility may be biological or chemical.

Biological Methods

The biological methods consist of raising a crop or a microbial culture in a field or in a sample of the soil, and estimating its fertility from the volume of crop or microbial count. Although these methods are direct estimates of soil fertility, they are time consuming and, therefore, not well adapted to the practice of soil testing. On the other hand, chemical testing of a soil is reasonable rapid and is therefore more handy for advisory work on a mass scale.

Chemical Methods

The chemical methods for evaluating the soil fertility consist of analyzing a sample of the soil for its content of essential plant nutrients which are expected to be in relatively short supply and whose deficiencies can be corrected by appropriate additions of suitable fertilizers. The approach to analyze a soil for the sample with strong acids or alkalis was given up when it was found to have a poor relationship with plant available quantities of the nutrient. The usual practice today is to obtain an extract from soil by adding to it a small volume of a solution containing mild chemical reagents. Soil is air dried and ground to pass a 2-mm sieve. The nutrient concentration is measured in an aliquot of the extract and results are expressed in terms of the quantity of that nutrient on per unit mass or volume basis. The choice of a chemical method to estimate nutrient availability in soils involves a selection of the most appropriate extractant and definition of the extracting procedure which may together remove quantities of the nutrient that are highly correlated with the biological index of nutrient availability on a group of soils. There is almost always some theoretical basis for the choice of extractant. Ranking of different chemical methods of soil testing depends upon the analysis of the selected group of soils by each of the methods, correcting the sets of estimates with the biological indices and listing them in the order of the coefficients of correlation. Multi-nutrient extractant: The analysis for more than one nutrient through a single extract simplifiers soil testing. A combination of equilibrant and extractant was proposed by Baker and Amacher (1981).Their solution contain 2.5×10^{-4} molar KCl, 10×10^{-4} molar $MgCl_2$ 50×10^{-4} molar $CaCl_2$ and 4×10^{-4} molar diethyl amine triamine penta acetic acid (DTPA), and is buffered at pH 7.3 with tri-ethanolamine. The proposed concentration of K, Mg and Ca chosen for the test solution represent optimum availability levels for the growth of most crops. The changes in the solution composition during equilibration with the soil provide information on the deviation of soil from the idea but these do not affect the quantities of micronutrients and trace elements brought into solution by the chelating agent, DTPA. If the concentration of K, Mg, or Ca in the solution after equilibration is equal to or greater than the initial concentration, the availability of the ions in the soil is considered adequate, Molar activities of phosphate are also obtained from the same extract. The phosphorus extracted by the Bray and Kurtz (0.025 N HCI+0.03 *N* NH_4F) extractant is considered an index of the quantity factor of that element. The electro-ultra filtration with water as the extractant was suggested by Nemeth (1979). A suspension of soil in water (1:10) is stirred in the central chamber of an assembly of three chambers, connected by membranes which permit the passage of solutes but not solids. A direct current is passed between electrodes in the outer chambers. It moves anions through the filter to the positive electrode where they form acids. The cations move through the other membrane to negative electrode to form hydroxides. Suction is applied to the outer cells to collect nutrients solutions at designated intervals for analysis.

Interpretation

The interpretation of soil test results requires a lot of background information and much intuitive thinking. It is useful to know the variation in soils with respect to pH, organic matter and soil texture and special problems like water logging or sodicity. Ordinarily there is a greater probability of obtaining a profitable response from fertilization on soils testing low in a nutrients element than soils testing high in that element. However, when difference in any one or more of the aforesaid soil attributes interpretation of soil test results becomes increasingly complex. One way of interpreting the soil test results is to classify them into low, medium and high categories. Such a classification is based on the observation, as already stated, that a low testing soil in the absence of an applied nutrient, produces a very poor crop and when a small amount of that nutrient is added to it, the resultant increase in yield is much more in this soil than what would come about from applying an identical amount of the nutrient to a soil which is high or even medium in its native supply.

Calibration

The work of determining the calibre of a soil test is known as Calibration. The calibration of soil tests can be represented by a linear equation:

$$Y = a + bx,$$

where, Y is the yield of a nutrient × or a crop yield, × is the nutrient element whose deficiency is corrected by applying a fertilizer, and b is the yield increase per unit quantity of the deficient nutrient. Depending upon the size of the experiment, more than one nutrients and soil test estimates from more than one soil depths can be related with crop yield in a single multiple regression equation.

Targeting Crop Yields-Considering the crop yield to be continuous function of plant nutrient supply in the growth medium, calibrations were obtained for different levels of soil fertility. The procedure of creating different levels of fertility artificially in adjacent plots was adopted to ensure homogeneity in soil management and weather whose diversity in experiments performed at different locations and in different seasons usually leads to poor correlations. From such experiments, the nutrients requirement (X) is estimated using following equation:

$$X_f = \frac{X_t\,Y - X_o X_s}{X_r} \qquad ....(2)$$

Where,

X_f = the fertilizer nutrient (kg/ha) to attain the target yield,

X_t = the amount of nutrient (in kg) in the grain and straw of the fertilized crop per mega gram of grain,

Y = the target grain yield in mega grams per hectare,

X_o = the amount of nutrient (in kg) in the grain and straw of the unfertilized crop per hectare per kg of soil test nutrient per hectare,

X_s = the soil test nutrients in the unfertilized control in kg per hectare.

X_r = the fractional recovery of fertilizer nutrient in grain and straw, averaged over treatments in the calibration experiments with different quantities of fertilizer nutrients.

Critical values are especially useful when public administration has high concern for increasing crop production in areas producing generally low to moderate yields and adequate profitability of a costly input is vital to the propagation of technology for improved agriculture A set of critical soil test values for use by soil testing laboratories in India as suggested by Muhr *et al.* (1965) are given in Table 1.

Table 1. Critical soil test values for available N, P, and K different

Nutrient	Low	Medium	High
1. Organic carbon (%)	>0.5	0.5-0.75	<0.75
2. Available N (kg/ha)	>280	280-560	<560
2. Olsen's P (P_2O_5 kg/ha)	>22	22-56	<56
3. NH_4Ac-K (K_2O kg/ha)	>110	110-280	<280

(Source: Muhr et al. 1965)

Soil test summaries

Soil testing provides sound information about the fertility and productivity of the soils. This enables the farmer to make the most profitable use of some of the costly inputs in farming. However, neither all farmers have the information or inclination to get all their soils tested nor is the soil testing service adequately equipped to perform this job on all the millions of fields across the length and breadth of the country. Hence, there is a need to compile soil wise and crop wise soil test summaries for all important agro-eco- regions, where already, there is sufficient use of fertilizers or where higher demands in the future are anticipation. It is necessary to have sufficient number of soil samples as well as representing sampling in order to ensure correct assessment of the prevalent situation. The soil test results are depicted in either of the two ways:

1. Categorizing areas having more than 50%, 25-50 or less than 25% nutrient-deficient soils, and
2. Nutrient index value calculated from the proportion of soils under low, medium and high available nutrient categories, as per represented by expression

$$NIV = \frac{N_1 + 2N_m + 3N_b}{N_1 + N_m + N_h}$$

Where, NIV is nutrient index value, N_1, N_m and N_h are the number of soil samples falling in the category of low, medium and high nutrient status and are given weight age of 1, 2 and 3 respectively.

2

Plant Nutrition

One hundred and nine elements have been identified and included in the periodic table so far. Most of this exit in the earth's mantle, earth's crust and soil. Before, moving further, let us understand some of the key term which is often used in this chapter.

Nutrient: A nutrient element is one that is required to complete the life cycle of the organism and its relative deficiency produces specific deficiency symptoms. Nutrients content is considered deficient when it is so low that it severely limits growth and produces characteristic deficiency symptoms. Range of nutrient content in plants associated with optimum crop yields is called sufficient. When the concentration of a nutrient element rises too high to cause significant growth reductions, it is termed as toxic.

Available nutrients: In the soil a nutrients elements is distributed in different discrete chemical forms, which often exist in a state of dynamic equilibrium and constitute the pool from which plants draw it. The nutrient available to the biological organism is termed as bio available nutrients. It is that portion of the nutrient in the soil that can be readily absorbed and assimilated by the plants. An available nutrient constitutes only a small portion of the total nutrient present in the soil.

Beneficial elements: Beneficial elements are the mineral elements which stimulate plant growth but are not essential or which are essential only for certain plant species, or under specific conditions like silicon, sodium, aluminium, cobalt, selenium and vanadium.

Fractional nutrient: This term introduced by Nicholas (1961) is defined as an element that plays a role in plant metabolism, whether or not that role is specific or indispensable.

Trace element: Trace elements are an element found in low concentration, perhaps less than one ppm or still less in soil plant and water, etc.

Tracer element: Radioisotope or a stable isotope of an element used for tracing its path in a system to study the mechanism of its interaction with the system is called a tracer element.

Heavy metal: A metal having specific gravity of more than 5.0 or having atomic number higher than 20 is termed as a heavy metal. As a corollary, any metal heavier than calcium is a heavier metal.

Nutrient content: Concentration of a nutrient or its amount per unit weight of a plant tissue is termed as nutrient content. Nutrient content is expressed in term of percent (kg/100kg or g/100 g) or ppm (parts per million) which is equivalent to mg of nutrient /kg of dry matter (mg/kg) or µg of nutrient per g of dry matter (µg/g). The percent (%) of a nutrient is converted to parts per million by multiplying with a factor of 10,000 or 10^4.

Nutrient Accumulation: Storage of a nutrient in a particular part or portion of the plant is called nutrient accumulation.

Nutrient uptake: Amount of nutrient taken up by the growing crops from either the soil or other sources is called nutrient uptake. Suppose roots, straw and grain of wheat crop producing dry matter of 2.5, 6.0 and 4 t/ha contain 0.2, 0.5 and 1.5 % N respectively, than the uptake of N will be as follows:

N uptake by roots	=2.5(t/ha) x 0.2 (kg/100 kg),	=0.005 t/ha	= 5.0 kg /ha.
N uptake by straw	= 6.0(t/ha) x.0.5 (kg/100 kg)	= 0.03 t/ha	= 30.0 kg /ha.
N uptake by grain	= 4.0(t/ha) x.1.5 (kg/100 kg)	= 0.0.6 t/ha	= 60.0 kg /ha.
Total uptake of N by Wheat	= 5.0+30.0+60.0 kg/ha	= 95.0 kg/ha	
Total N uptake by the above ground portion		= 30+60	= 90.0 kg/ha.

Nutrient Removal: The nutrient contained in the harvested portion of the crop is termed as the nutrient removed.

Essential nutrient elements

The chemical compounds/ ions required by an organism are termed as nutrients and their supply and absorption for growth is defined as nutrition. The higher plants require solely inorganic nutrients essential for their growth and development and fundamentally differ from animal human beings and number of microorganisms which also need organic food stuffs in addition to inorganic chemical nutrients.

The following three criteria must be met to consider a nutrient element essential for the plants.

(1) A deficiency or absence of the element should make it impossible for the plant to complete its life cycle.
(2) The deficiency should be specific for the element in question
(3) The elements should be directly involved in the nutrition of the plant or plays a direct role in its metabolism e.g. as a constituent of an enzyme system.

An element is said to be essential if the plant can not complete its life cycle without it. Out of the 107 elements in the Periodic Table, 90 elements naturally occur in plants, out of which only 17 are essential nutrients. In the absence of one essential nutrient, the presence of all other essential nutrients becomes meaningless because the plants will not be able to utilize these nutrients for its growth and development. The essential nutrient elements are carbon, hydrogen, oxygen, nitrogen, phosphorus, potassium, calcium, magnesium, sulphur, iron; manganese, zinc, copper, boron molybdenum, chlorine, and nickel. These are recognized as universally essential nutrient elements, because these meet the essential requirements for diverse groups of organism algae bacteria fungi and green plants. Cobalt has been established to be essential for leguminous plants only.

Plants nutrients: functions and deficiency systems

Plants require 17 essential nutrients for their normal growth and completion of life cycle. On the basis of their relative concentration or amount in plant tissue, these are divided into (i) macronutrients and (ii) micronutrients.

Macronutrients

These are required in large amount (g/kg dry matter) by the plants and hence are termed as macronutrients. These are C, H, O, N, P, K, Ca, Mg, and S. Earlier these were also known as major nutrients. Out of these N, P, and K are the primary nutrients and Ca, Mg and S the secondary nutrients and the rest are micronutrient elements. This classification is based on their relative abundance and prevalence of deficiency and not on their relative importance in plant nutrients.

The deficiency symptoms are generally characteristics and appear on a specific part of plants i.e. leaves, stem and roots depending upon its mobility and role in the plants. The deficiency symptoms are used for assessing the fertility of soil and for correcting the nutrient deficiency in question directly in the field without performing costly soil or plant analysis. Hence the specific and general functions and visual deficiency symptoms of the essential plant nutrients are described to help better understanding the fertilizer or nutritional needs of plant vis-à-vis soil fertility status.

Carbon oxygen and hydrogen: Plants take up C, H, and O (about 90-95% of the plant composition) mainly from the air and water. The C and O are taken up from the atmosphere and possibly as HCO_3 from the soil solution. Further oxygen is also taken up partly by plants in molecular form (O_2). Hydrogen is taken up as water from the soil solution and also from atmosphere and photolysis.

Function: C, H and O are constituents of organic materials are involved in enzymatic processes and oxidation –reduction process.

Deficiency symptoms: The deficiency of C, H and O in plants is rare because these nutrients occur in abundance in the nature The exceptions are extreme conditions of water deficit and water logging under which the uptake of H and O is severely restricted in wilting due to water deficit and yellowing of lower leaves of plants because of O deficiency, which in turn also cause N deficiency and ultimately the death of plants.

Nitrogen: plants absorb N from the soil solution as nitrate (NO_3) and as ammonium (NH_4) e.g. rice. The content of N in healthy plants ranges between 1 and 5% depending upon the species or variety. Nearly 62% soils are deficient in N.

Function: Nitrogen plays an important role in plant metabolism. As a constituent of chlorophyll it harnesses solar energy and fixes atmosphere CO_2 as carbohydrates. The plant gets stunted and develops chlorosis. Excess of N produces leathery dark green succulent leaves or crop which delays maturity and increases susceptibility to disease and lodging. Generally, application of 120-180 kg N /ha to irrigated wheat, rice, maize and 240 N kg/ha to sugarcane and rapeseed-mustard in various soil.

Deficiency symptoms: plants containing less than 1.0% N are generally deficient in N. Symptoms of N deficiency first appear on the older leaves. The N deficient plants are generally stunted develops late and produces less flowers as well as shriveled grains.

Phosphorus: The content of P in normal plants ranges from 0.1% to 0.4%. Soil is the main source of P for plant nutrients. Plant root absorb largely as dihydrogen orthophosphate ($H_2PO_4^-$) ions and under neutral to alkaline condition as orthophosphate (H_2PO_4–) ions from the soil solution. Further, the plant also absorbs very small amounts of solution organic phosphates *viz.* nucleic acids and phytin. It should preferably be applied at sowing.

Functions: phosphorus plays an important role in energy transformations and metabolic processes in plants. It stimulates plant growth. It is necessary for cell division, merismatic growth root, and seed and fruit development as well as in stimulating flowering, ear emergence and maturation of crop.

Deficiency symptom: Crop grown on acid soils, calcareous soils, coarse textured soil low in organic matter suffer from P deficiency. In P deficient soil soils the plants develop visual deficiency symptoms, *viz.* they fail to make a quick start, develop poor root system, and remain stunted; being mobile nutrient in plant. P need of crops to produce optimum yield on soil generally 50-60 kg P_2O_5/ ha for wheat, 30-40 kg P_2O_5/ha for each crop in rice–wheat system, 20-40 kg P_2O_5/ha for sunflower, safflower and linseed.

Potassium: Potassium absorbs K as K^+ ion in large amounts from the soil than any other nutrient (except N and Ca). Plants can store K in large quantities than what is needed for optimum growth without causing toxicity which is termed as luxury consumption. Potassium is deficient in 20% soils of India.

Functions: Potassium is involved in regulating the opening and closing of the stomata, (which is the essential for photosynthesis) water and nutrients transport and cooling of plants. It activates a number of enzymes. K helps in maintaining cytoplasmic pH to 7 and, ideal for many enzyme reactions; it activates nearly 60 enzymes. It imparts resistance to plants against fungal and bacterial diseases.

Deficiency symptoms: Potassium is highly mobile in plants and its deficiency causes on visual symptoms on older leaves. The symptoms of chlorosis start from the leaf margins, followed by scorching and browning of tips and margin in potato, wheat, oat and maize. Generally 30-40 kg K_2O/ ha as basal dressing for irrigated rice wheat, maize etc alleviate its deficiency.

Calcium: Plant roots absorb calcium as Ca^+ ions and its content in healthy plants ranges from 0.2 to 1.0%.

Function: It is a constituent of the cell wall, an activator of different plants enzymes and is essential for the stability of cell membrane. As calcium is immobile in the plant hence, typical visual deficiency symptoms appear on the younger leaves. Encourage seed production. Improves intake of other plant nutrients like N and trace elements such as Fe, B, Zn, Cu, and Mn by correcting soil pH.

Deficiency symptoms: Its deficiency is not common. But in acid soil because of its leaching losses and in alkali or sodic soils due to excess Na, the deficiency constraints of Ca are encountered. Typical visual deficiency symptoms appear on the younger leaves. The Ca-deficiency syndromes in apples are bitter pit and in tomatoes blossom rot, end rots. Lime application to acid soils and gypsum to alkali or sodic soils correct Ca- deficiency.

Magnesium: Plants roots absorb magnesium as Mg^{2+}ion and its content in Mg adequate plants ranges from 0.1to 0.4% below which the deficiency may occur.

Function: It is an essential constituent of the chlorophyll as Mg-porphytin. It regulates the activity of several enzyme systems involved in synthesis of nucleic acids and metabolism of carbohydrates. A large part of magnesium is associated with organic ions *viz.* malate and it also help in the movement of sugars and translocation of P in plants.

Deficiency symptoms: Magnesium deficiency is also observed in plants growing in acid soils, leached soils and sandy soils. As Mg is a mobile nutrient in plants, its deficiency symptoms Manifest first on the older leaves and advances upwards to younger leaves as interveinal chlorosis followed by development of purple lesions within the chloratic and the tissues. As deficiency proceeds, the veins also become chlorotic and the leaves develop almost uniform pale colour and the purple lesion turn brown orange or red: ultimately the tissues dry up and may die.

Sulphur: plants absorb Sulphur largely as SO_4^{2-} ion and as SO_2 by the foliage. The concentration of S in healthy plants ranges from 0.1 to 0.4% almost equal to that of P and Mg.

Function: It is an essential constituent of the Sulphur-containing amino acids, viz. cysteine, cystine and methionine. It also a constituent of ferrodoxin-containing nitrogenase enzyme, which is involved in biological N fixation (BNF) and electron transfer reaction and in the metabolic activities of vitamins, biotin, thiamine, coenzyme. A synthesis of glucosides in mustard oil, and in increasing the oil quality of oil seeds crops.

Deficiency symptoms: Generally crops grown on coarse textured soils under intensive cultivation suffer from its deficiency, especially fertilized with high analysis fertilizer free from S and /or irrigated with tube well and canal water low in S or in rainfed areas. Soil low in organic matter and coarse textured or leached acid soil are deficient in S. As sulphur is immobile in the plants its deficiency first shows up on the younger leaves as uniform pale green followed by their chlorosis. Depending upon the crop and variety plants having less than 0.1 to 0.2 % S contents suffer from its deficiency. Crop plants having more than 16:1 N:S ratio can be suspected to be deficient in sulphur. Application of 20-60kg S /ha corrected its deficiency in most of these soils/crops.

Micronutrients

These are required in small or micro amount (<50mg /kg dry matter except Fe and Mn) and hence are termed micronutrients. These are as essential and important like macronutrients and include Fe, Mn, Cu, B, Mo, and Ni. These are subdivided into micronutrient cations (Fe, Mn, Zn, Cu, Ni) and micronutrients anions (B, Mo, and Cl)

Iron: Iron exhibit Fe (II) and Fe(III) oxidation states in plants, and is taken up in the form of Fe^{2+} ion. Its concentration in normal plants ranges between 100 and 50 mg/kg.

Function: It is a constituent of haeme and non heame proteins. These are involved in redox reactions in respiration and photosynthesis. It is necessary for the synthesis and maintenance of chlorophyll in plants and plays an important role in nucleic acid synthesis. Iron activates a lot of enzymes.

Deficiency symptoms: Iron deficiency is observed in plants growing on calcareous high pH soils. Iron is immobile in plants usually 30mg Fe^{2+}/kg dry matter is considered the critical level for Fe deficiency. The iron deficiency shows first on younger leaves as pale yellow interveinal chlorosis of the tissues. Generally the plants containing less than 50 mg Fe^{2+} /kg dry matter are categorized deficient in iron. In coarse textured soils organic manures including green manures along with foliar sprays of Fe are more effective in correcting Fe deficiency.

Manganese: Plants absorb Mn as Mn^{2+}, but it is easily oxidized to Mn^{3+} and Mn^{4+} form. The common Mn carriers that can supply Mn are inorganic salts such as manganese sulphate which is recognized as fertilizer in the FCO.

Function: it is a constituent of water splitting enzyme associated with O_2 evolution. Manganese is an integral component of the water splitting enzyme associated with photo system II. It is a constituent of super oxide dismutase (Mn-SOD). Manganese has a role in tricarboxylic acid.

Deficiency symptoms: Manganese-deficient plants contain less than 25 ppm Mn. Deficiency symptoms of Mn are more severe on middle leaves than on the younger one. Deficiency symptoms are also observed on plants grown in calcareous soil. In the monocotyledonous plants like cereals deficiency symptoms of Mn appear as greenish grey spot. The syndromes of Mn deficiency are popularly known as : grey speck of oats, speckled yellow of sugabeet, marsh spot of peas, *pahala* blight of sugarcane.

Zinc: Plants absorb Zn as zinc ions (Zn^{2+}). Zinc sufficient plants contain 27 to 150ppm Zn in mature tissues. Since it does not have variable valency and occurs only as Zn(II) oxidation states.

Function: Zinc is a constituent of three enzymes like carbonic anhydrase (CA), alcoholic dehydrogenase (AD) and super oxide dismutase (SOD). It is involved in the synthesis of indole acetic acid and proteins. Zn plays an important role in the stabilization and structural orientation of the membrane proteins. Zn influences translocation and transport of P in plants.

Deficiency symptoms: Plants containing less than 15 ppm Zn are regarded deficient in plants. Common deficiency symptoms of Zn are interveinal chlorosis, first appearing on the young leaves, and reduction in the size of leaves. This deficiency is observed in crop growing in soils which are coarse in texture high in pH (sodic), high in $CaCO_3$ (calcareous) or low in organic matter or water logging / submerged rice soils. A white to pale yellow chlorosis develop between the mid vein and margin and extent towards the tip, popularly known as white bud in maize or sorghum. The white bud is not a bud but rather a bend or patch of white tissue. Name given to Zn deficiency symptoms include: Khaira disease of rice reported from India (also called Hadda disease Pakistan, Akagare type II in Japan, Taya-taya and Apayaopula in Philippines), white bud of maize, mottle leaf (little leaf) or frenching of citrus and little leaf of cotton.

Copper: It exits in plant as cuprous Cu (I) and cupric (II) oxidation states and is taken by plants as Cu^{2+}. Copper is a transition element exiting in the plants as a component of a large number of proteins and enzymes as Cu (II) and can exchange its valency as per equilibrium:

$$Cu^{2+} + e^{-} \rightleftharpoons Cu^{+}$$

Function: Copper is a constituent of number of enzymes or proteins that perform important biochemical function. Copper is important in imparting disease resistance to the plants. It enhances the fertility of male flowers.

Deficiency symptoms*:* Plants having less than 5 ppm Cu are regarded as Cu-deficient. Male flowers sterility, delayed flowering and senescence are the most important effects of Cu-deficiency. Cu-deficiency is generally observed in crops growing on soils of either inherently low in total Cu or soils high in organic matter (peat soils). The plants contain concentration ranging between 5 and 20 mg/kg. The copper is immobile in plants, its deficiency symptoms appear on the young mature leaf and turn pale green with interveinal pale yellow chlorosis at the base of the blade.

Molybdenum: Mo is the only heavy transition metal taken by the plants as molybdate ions ($MOO^{2-}{}_{4}$). In the plant system it exists as Mo(VI) under oxidative conditions and undergoes reduction to Mo as Mo(V) and Mo(IV) forms.

Function: Mo is an essential constituent of many enzymes: nitrogenase that catalyzes biological nitrogen fixation (BNF) by root nodules bacteria of legumes plants. Nitrate reductase (NR) reduces nitrate to nitrite and xanthine oxidase /dehydrogenase which catalyzes the catabolic pathway of purines to uric acid. Mo affects the formation and viability of pollens.

Deficiency symptoms: Mo deficiency is observed in plants growing in acid soils. Mo deficiency plants usually contain less than 0.2 mg Mo/kg and the sufficient ones 0.2 to 2 mg mo/kg. Its deficiency symptoms appear on old or middle leaves all over surface. Mo deficiency in cauliflower and broccoli is termed as whip-tail.

Boron: Boron is absorbed by the plants mainly as boric acid (H_3BO_3) and high pH conditions also as H_2BO_3. Most important property of boron is to form stable complexes with organic compounds.

Function: It is responsible for the cell wall formation and stabilization lignifications and xylem differentiation. It imparts drought tolerance to the crop. Boron plays a role in pollen germination and pollen tube growth. It facilitates ion uptake by way of increasing the activities of plasma –membrane bound H^{+} -ATPase. It facilitates transport of K in guard cells as well as stomal opening.

Deficiency symptoms: Boron is immobile in plant, and hence its deficiency symptoms develop on new tissue or young leaves as white or transparent lesion in the interveinal mid areas of the leaf. B deficiency in crops is more critical in highly calcareous, sandy, leached soils, limed acid soils. Its critical limits range between 20 and 70 mg/kg in dicotyledonous and 5-10 mg/kg in graminaceous plants. The deficiency syndromes are popularly known as brown heart or heart rot in sugar beet and turnip, browning of curds and hollow stem in cauliflower and external and internal cork in apples.

Nickel: Nickel is absorbed by plants as nickel ions (Ni^{2+}). Normal plants contain 0.1 to 10 mg Ni /kg.

Function: Nickel is associated with nitrogen metabolism. It is needed for grain filling and seed vitality. In free living Rhizobia, adequate Ni supply ensures optimum hydrogenase activity. It facilitates transport of nutrients to the seed or grain.

Deficiency symptoms: Critical level of Ni deficiency in barley shoots is 0.1 ppm. Its deficiency causes accumulation of nitrates and decrease in amino acid content in barley, containing less than 0.1 mg Ni/kg.

Chlorine: Chlorine is absorbed as chloride ions (Cl^-) by the plants. Normal plants have ranging from 100-500 ppm (mg Cl/kg).because of its abundance in nature, its deficiency is noticed only in areas far from the sea where atmosphere deposition does not supply enough Cl and its deficiency have not been reported from any where in India.

Function: It plays a major role in osmoregulation and charge compensation in higher plants. It acts as a cofactor in Mg containing water splitting enzyme of photo system II. Cl in abundance suppresses the plant disease. Chlorine supply improves the nutritional quality of vegetables by preferential lowering the NO_3 –N concentration in tissue.

Deficiency symptoms: Plants having less than 100 ppm Cl is usually designated as deficient. The symptoms appear as irregular yellow to necrotic spotting of leaves, premature wilting, and chlorosis of newly emerging leaves and reduced shoot and root growth.

Identification of deficiency: A brief key to deficiency symptoms developed by Finck (1992) is give below. It is also desirable to understand the role of such factors that might produce symptoms similar to other of the nutrients.

Symptoms appear first on older leaves

Nutrients	Deficiency symptoms
Nitrogen	Chlorosis starting from leaf tips
Phosphorus	Reddish colour on green leaves or stem
Potassium	Necrosis on leaf margins
Magnesium	Chlorosis mainly between veins
Manganese	Brownish grayish whitish spot

Symptoms appear first on younger leaves

Sulphur	Mottled yellow green leaves with yellowish veins
Iron	Mottled yellow green leaves with green veins
Manganese	Brownish black spot
Copper	Youngest leaf has white tip
Boron	Youngest leaf is brownish or dead

Beneficial elements

Sodium: Sodium accumulates preferential in the vacuoles and plays a role in maintaining the solute potential of the cell. Sodium can replace some of the essential functions of K in plants. It improves the water balance of plants under limited water supply or under arid climate.

Silicon: Silicon has beneficial role in rice and sugarcane crops. It contributes to rigidity and strengthening of the cell wall. It enhances the physiological availability of zinc in plants and counteracts zinc- deficiency – induced phosphorus toxicity.

Cobalt: Cobalt enhances N_2 fixation in legumes and improves the nutritional quality of the forage crops for ruminants.

Selenium: The chemistry of Se is similar to that of S with Se existing in Se^{2+}, Se^{4+} and Se^{6+} states. Se is an essential element for animals. To avoid Se- deficiency in animals the concentration of se in diets should be 0.1 to 0.3 mg/kg dry matter. In livestock symptoms of Se- deficiency include white muscle disease or nutritional muscular dystrophy.

Aluminium: In plants species having high tolerance and capacity for Al, the beneficial effects of Al on growth accrue from its ability to alleviate toxicity of P, Zn and Cu.

Vanadium: Vanadium (V) favours nitrogen fixation in leguminous plants. The chemical behavior of V and Mo is similar. It is brought that Vanadium might partly substitute the function of Mo in nitrogenase system of *Azotobacter*. Other elements being investigated for the beneficial effects on plants including Bromine, iodine, lead, cadmium, chromium, fluorine etc. but no evidence has been found so far.

Nutrient Movement in soil: Transpiration of water by plants induces the movement of water from soil to plants. Nutrients which are dissolved in soil solution move to the plant roots with water in response to the hydraulic gradient by created by the roots as a result of continuous absorption of water. However, differential requirements of nutrients vis-à-vis water by plants result in either accumulation or depletion of nutrients in the vicinity of root surface. There are two separate processes involved in the movement of nutrients to plant roots. Movement of nutrients with water is known as 'mass flow' and the movement through water as 'diffusion'. In addition, growing plants roots move into the soil and displace it. In the process, roots come in contact with the soil and intercept the nutrients from the soils through contact exchange. Thus, three mechanisms recognized for nutrient movement are–mass flow, diffusion and root inception.

Mass Flow: Mass flow is the movement of nutrients through the soil to the root in the convective flow of water caused by the plant water absorption. The nutrients move physically with the flow of soil water towards plant root, because continuous absorption of water due to transpiration creates a hydraulic gradient at the root surface, where their absorption through the roots takes place along with water. Some nutrients ions that move freely with water, generally for large distances, are called mobile nutrients. However, others move slowly with the water for relatively small distance, only a few millimeters, called immobile nutrients. The nutrient uptake through this mechanism is directly related to transpiration-the amount of water used by plants. Maas flow is responsible for supplying the root with much of the plant needs for N, S, Cl, Ca, and Mg when presents in sufficient concentration in the soil solution, but not for relatively immobile nutrient elements e.g. P, K, Zn etc. If a crop producing a dry matter of 10 t/ ha uses 300 ml of water to produce one gram of dry matter during crop growth, given the average concentration of Ca10^{-2}:K,10^{-3} and 10^{-4} for P, 10^{-5} to 10^{-6} moles per litre. Which amount of water would carry and which would carry about Ca K and P?

Solution: total dry matter produce = 10 t/ha= 10 X 10^3 kg/ ha = 10^4 kg/ha X 10^3 kg/ ha = 10^7 g/ ha

Amount of water used for producing one gram of DM = 300 ml

Amount of water used for producing of 10^7 g/ ha DM= 10^7 X300 ml

= 3 X 10^9 ml

= 3X 10^6 L

Concentration of Ca in soil solution = 10^{-2} M/L =10^{-2}X 40 = 4x10^{-1} g/l

Amount of Ca moved along with water to plant ==3X 10^6 L X 4x10^{-1} g/ l= 12 X 10 ^{5} g per litre or 12 X 102 kg =1200 kg. Since, amount of water transpired is for one hectare, hence Ca moved by mass flow is 1200 kg /ha.

Diffusion: When mass flow is unable to supply sufficient quantities of a nutrient and a continued uptake occurs, the concentration of the nutrients at the root surface is reduced and a concentration gradient is established. Ions move by Brownian movement from points of higher concentration to those of lower concentration when a concentration gradient exists. Movement of a nutrient ion in response to the concentration gradient is termed as diffusion.

Root interception: The term root interception coined by Stanley A. barber of Purdue State University, USA, is used to describe the soil nutrients at the root surface that do not have to move to the interface to be positionally available for absorption but are approached by the roots itself in the soil. The quantity of the nutrients supplied by root-interception is taken as the quantity present in a volume of soil equal to the root volume. It is generalized that root volume for several crop species growing in soil is less than one percent of the soil volume in 0-20 cm depth. Hence, calculations are based on the volume. Conceptually, as a root system develops and exploits more soil, soil solution and soil surface retaining absorbed ions are exposed to the root mass and adsorption of these ions occurs by a contact exchange mechanism of Jenny and Overstreat (1939). Root interception of the nutrients in fertile soils can be enhanced by the mycorrhiza, a symbiotic association between certain fungi and plant roots.

3

Fertilizers

Fertilizers usually referred to as chemical fertilizers or commercial fertilizers are the synthetic or naturally occurring chemical compounds applied to soil to supply plant nutrients for crop growth. Unlike in manures, the nutrient elements in fertilizers are present in higher concentration and in forms which can be readily utilized by plants directly or after rapid transformation. Being concentrated in nutrient elements, the fertilizers have the advantage of smaller bulk, economy and ease of transport, storage and handling. Moreover, their dose can be adjusted to suit the requirement as determined by soil fertility evaluation.

The high nutrient content in easily available form has led to the world wide adoption of fertilizer use for increased and sustained crop production and is now the key factor in crop production. That the proper use of fertilizer greatly increases crop production is well established. However, proper benefits from fertilizer use can not be realized without the use of improved varieties of crops, better cultural practices, and plant and crop management. It is a mistake to thinks that fertilizers are the substitutes for manures which have their role to play. The use of fertilizers makes it possible to introduce extra supplies of nutrients into cycles of growth and decay, thereby improving soil fertility. Fertilizers are inorganic materials which can supply plant nutrient in available form, having the high analytical value and having a definite composition and mostly are industrial products.

CLASSIFICATION

Some of the fertilizers supplies only ONE primary nutrient element. Such fertilizers are ammonium sulphate that supplies nitrogen (but no P and K), super phosphate that supplies phosphorus (but no N and K), and muriate to potash that supplies potassium (but no N and P). They are termed as STRAIGHT FERTLIZERS. Some of the fertilizers supplies MORE THAN ONE primary nutrient element. Such fertilizers are ammonium phosphate that supplies nitrogen, phosphorus, nitric (or, nitro) phosphate (15-15-15) that supplies N, P, and K. They are termed COMPLEX OR MIXED FERTILIZERS. Some times, they are mixed products. *e.g.* nitric phosphate(20-20-0) is mechanically mixed with muriate of potash to obtain nitric phosphate (15-15-15).

Some time these fertilizers are not mixed products. Each of them contains only ONE chemical compound that supplies more than one primary nutrient element. For example, mono ammonium phosphate compounds that supplies nitrogen and phosphorus. Mono ammonium phosphate, diammonium phosphate, urea ammonium phosphate, nitric phosphate (20-20-0) supply only one two primary nutrient elements, *viz.*, nitrogen and phosphorus (but not all the three primary nutrient elements). They are termed incomplete complex (or mixed) fertilizers Nitro or nitric phosphate (15-15-15), and different N-P-K fertilizers supply all the THREE primary nutrients elements. They are termed COMPLETE mixed fertilizers.

Some of the straight fertilizers *viz.*, ammonium sulphate urea, ammonium sulphate nitrate, calcium ammonium nitrate, sodium nitrate supply nitrogen. Hence, they are termed NITROGENEOUS FERTLIZERS or NITROGEN FERTLIZERS or shortly N FERTLIZERS.

Some of the straight fertilizers, *viz.*, super phosphate supply phosphorus. Hence they are called PHOSPHATIC FERTLIZERS or PHOSPHORUS FERTLTIZERS or shortly P FERTLTZERS.

Some of the straight fertilizers, such as muriate of potash, potassium sulphate etc supply potassium hence, they are termed as POTASSIC FERTLIZERS or POTASH FERTLIZERS or shortly K FERTLIZERS. The forms of nitrogen in ammonium sulphate, sodium nitrate calcium ammonium nitrate and urea are ammonium nitrate, ammonium plus nitrate and amide respectively. Hence, N fertilizers are respectively termed ammonium fertilizers or nitrate fertilizers or ammonium plus nitrate fertilizers or amide fertilizers. The forms of phosphorus in super phosphate ammonium poly phosphate and ammonium metaphosphate are orthophosphate polyphosphate and metaphosphate respectively. Hence they are respectively termed orthophosphate fertilizers or poly phosphate fertilizers and metaphosphate fertilizers. Again, the phosphate in super phosphate and all ammonium phosphate is soluble in water. Hence, they are termed water soluble fertilizers. The phosphate in basic slag is not soluble in water but soluble in neutral one normal ammonium citrate solution. Hence, it is termed CITRATE SOLUBLE P FERTILIZER. Citrate soluble phosphate ($CaHPO_4.2H_2O$) is significantly present in nitric phosphate. The phosphate in rock phosphate (apatite) is not soluble in water and neutral one normal ammonium citrate solution. Hence; it is termed INSOLUBLE P FERTILIZERS. The ammonium or amide containing N fertilizers, when added to soil, induced acidity hence, they are termed PHYSIOLOGICALLY ACIDIC FERTLIZERS.

Some of the fertilizers viz., CAN when added to soil does not change the soil reaction. Hence, they are termed PHYSIOLOGICALLY NEUTRAL FERTLIZERS. Sodium nitrate, when added to soil, induces alkalinity, through it is a NEUTRAL salt. Hence, it is termed PHYSIOLOGICALLY ALKALINE (BASIC) FERTLIZERS.

Nitrogenous Fertilizers

All those fertilizers, which contain excess of nitrogen and are prepared in factories on commercial scale, are called Nitrogenous fertilizers. Nitrogenous fertilizers may be classified into four groups on the basis of the chemical form in which nitrogen is combined with other elements.

1. **Nitrate fertilizers:** Nitrogen is combined as NO_3 or nitrate with other elements. Such fertilizers are:
   - *(i)* Sodium nitrate or Chilean nitrate ($NaNO_3$) – 16 % N
   - *(ii)* Calcium nitrate [$Ca(NO_3)_2$] – 15.5% N
   - *(iii)* Potassium nitrate(KNO_3) – 15 % N
2. **Ammonium fertilizers:** In these fertilizers, nitrogen is combined in ammonium (NH_4) form with other elements. Such fertilizers are:
   - *(i)* Ammonium sulphate [$(NH_4)_2$ SO_4] - 20% N + 24 % S
   - *(ii)* Ammonium phosphate ($NH_4H_2PO_4$) - 20 % N + 20% P_2O_5 OR - 20 % N + 20% P_2O_5
   - *(iii)* Ammonium chloride (NH_4Cl) - 24 to 26 % N
   - *(iv)* Anhydrous ammonia - 82 % N
   - *(v)* Ammonia solution - 20-25 % N
   - *(vi)* Diammonium phosphate - [$(NH_4)_2SO_4$]- 18 % N
3. **Nitrate and Ammonium fertilizers:** These fertilizers contain nitrogen in the form of both nitrate and ammonium. Such fertilizers are:
   - *(i)* Ammonium nitrate(NH_4NO_3) - 33% to 34 % N
   - *(ii)* Ammonium sulphate nitrate - 26% N
   - *(iii)* Calcium ammonium nitrate - 20.5 or 26 % N

4. **Nitrogenous solution:** These fertilizers found in liquid forms and marked in other country. Such fertilizers are:
   - *(i)* Aqua ammonia -20% N
   - *(ii)* Ammonia urea solution -35-36% N
   - *(iii)* Ammonia –ammonium nitrate soilution -35-50% N
   - *(iv)* Ammonia –ammonium nitrate-urea soilution -37-49% N
   - *(v)* Non-Ammonia –ammonium nitrate-urea soilution -32 % N
5. **Amide fertilizers:** These fertilizers containing in amide form. Such fertilizers are:
   - (i) Urea [CO $(NH_2)_2$] -46%N
   - (ii) Calcium cynanide ($CaNCN_2$) – 21% N
6. **Slow release nitrogenous fertilizers:** Slow /controlled release and stabilized synthetic fertilizers are grouped as follows:
   - *(i)* Urea form
   - *(ii)* Mixed urea –Neem, Karanj, Mahua, Pyrite, Rock phosphate, Clay etc.
   - *(iii)* Sulphur coated urea
   - *(iv)* Metal ammonium phosphate
   - *(v)* Guanal urea
   - *(vi)* Oxide
   - *(vii)* Crotonylidene diurea
   - *(viii)* Isobutylidene diurea
   - *(ix)* N-lignin

Manufacturing process and general properties of nitrogenous fertilizers:

General diagram of manufacturing main nitrogenous fertilizers is given below:

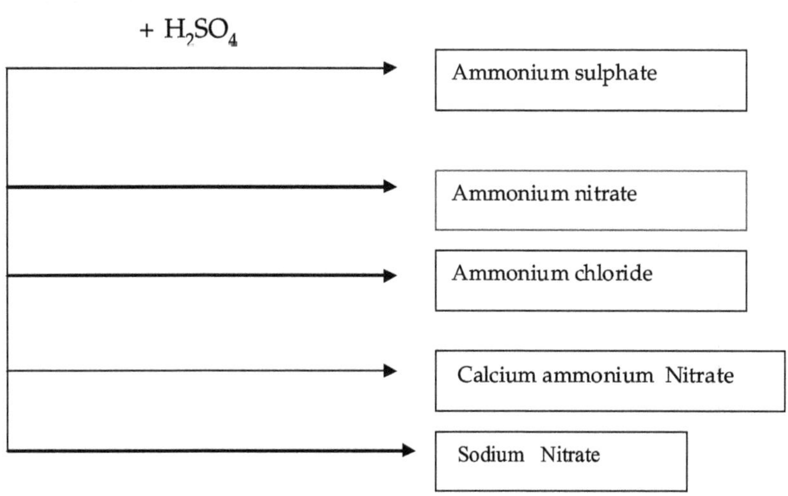

$$N_2 + 3H_2 \xrightarrow[\text{400-500 °C, 300 Atm. Pressures}]{\text{Promoter Mo/Al/K/CaO, } Fe_3O_4 + FeO} 2\,NH_3$$

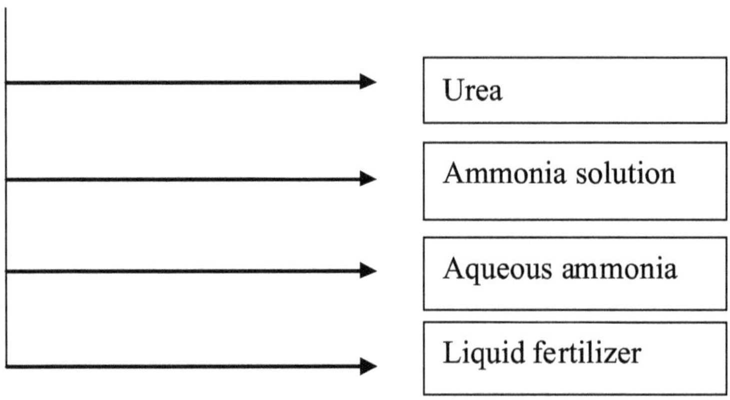

Ammonia is prepared by HARBER METHOD. Nitrogen reacts with hydrogen at very high pressure and temperature in the presence of catalyst mixture (Fe_3O_4+FeO). Main source of N and H_2 are coal, coal gas neptha, air and water. Once ammonia is formed, different nitrogenous fertilizers are prepared

Nitrate fertilizers

Sodium nitrate ($NaNO_3$) – It is also called Chilean nitrate or Chilean saltpeter or chilli salt petre because its deposit was first found in chilli country of South Africa. Now a-days its use is very limited.

Manufacturing: It is manufactured by treating sodium carbonate with nitric acid.

$$Na_2CO_3 + 2HNO_3 \longrightarrow 2NaNO_3 + H_2O + CO_2$$

Nitric acid is formed by the oxidation of ammonia

$$4NH_3 + 5O_2 \longrightarrow 4NO + 6H_2O + \text{Energy}$$

$$2NO + O_2 \longrightarrow 2NO_2$$

$$3NO_2 + H_2O \longrightarrow 2NHO_3 + NO$$

Properties: It contains 16 % N in nitrate forms. It also contains traces of boron and iodine. It is immediately available to growing plants. Since it is readily soluble in water, it is subjected loss by leaching. It is used exclusively as top- dressing. It is to be applied during the rainy season at sowing time and should be carefully drilled along with the seed. In addition to its nitrogen, commercial sodium nitrate contains approximately 27 % sodium. It is a basic fertilizer. Important properties are as below.

Colour	- White
State	- Granular Fertilizer, Crystalline (Pure)
Equivalent alkalinity	- 29.0
Relative density	- 2.26
Molecular weight	- 85.01
Melting point	- 308.3°C
Water Solubility at 100° C	- 150 g 100 ML^{-1}

Refine caliches by shanks process and Guggenheim process, sodium nitrate was also produced.

- Common salt and nitric acid (70%) are boiled to yield sodium nitrate.

  $3NaCl + 4HNO_3 = 3\ NaNO_3 + 2H_2O + Cl_2 + NaCl$

- Nitrogen is reacted with hydrogen under temperature (400-500°C) and pressure (200-1000 atm) is presence of catalyst (Fe_3O_4+FeO) and promoter (Mo/Al/K/CaO)

  $N_2 + 3H_2 \rightarrow 2\ NH_3$

- Ammonia is reacted with sodium carbonate solution to yield sodium nitrate. The reactions that occur is probably as follows.

  $4NO_2 + 2H_2O + O_2 + Na_2CO_3 \rightarrow 4\ NaNO_3 + 2H_2O + CO_2$

Sodium nitrate solids are obtained by evaporation and crystallization of the reaction product.

Effect on soil

Useful effect on soil acidity: Sodium nitrate reduces the soil acidity. Nitrate is taken up by plants and sodium is left is soil which forms NaOH reacting with H_2O. NaOH, a strong alkali reduces soil acidity and Ca and Mg, which are leached down in acidic condition, remain in soil. Thus, it can be successful be applied in acidic soil.

Bad effect on soil structure: deflocculating occurs due to excess use of sodium nitrate which causes big clods in clayey soil. The liberated Na^+ ions from $NaNO_3$ react with carbonic acid (H_2CO_3) and form Na_2CO_3. produces alkalinity in soil which disperses colloids and deflocculated structure is formed.

Effect on soil fertility: Soil fertility is adversely affected by the excess use of $NaNO_3$ because Na^+ ions replace most of ions present on soil colloids which causes deficiency of nutrients.

Effects on crops:

$NaNO_3$ should not be used in the area of excess of rain fall or in paddy crop because NO_3^- which is very much soluble in water, leaches down and become unavailable to plants. It can successful be used in maize, sugarcane, wheat barley, cotton etc. It is very good fertilizers for sugar beet and spinach.

Time and method of application

$NaNO_3$ should be mixed with other phosphatic and potassic fertilizers at the time of application in soil. It should not be added at a time. It should not be top-dressed because due to its high alkalinity. It should be applied with seed or near plant roots as side dressing. It should be applied in light neutral to acidic soil where water drainage is good.

Calcium Nitrate: [Ca $(NO_3)_2$]: It is also called nitrate of lime. It is first synthetic nitrogenous fertilizer. It use and manufacture is very limited in India.

Manufacture: It is manufactured by two processes is given below.

1 Nitrophosphate process: When rock phosphate is treated with HNO_3, Calcium phosphate, and calcium nitrate is formed. Calcium nitrate is separated by acidulation method.

2. Direct neutralization process: calcium nitrate is manufactured by neutralizing HNO_3 with $CaCO_3$.

   $CaCO_3 + 2HNO_3 \rightarrow Ca\ (NO_3)_2 + H_2O + CO_2$ + Energy

The 80% of total product by above reaction is solid in which 40% is calcium nitrate. 20 % ammonium nitrate is added to increase the N% 15.5.

Properties: It contains 15 % nitrogen. Important properties are as given below.

Colour	-White
State	-Granular Fertilizer
Relative density	- 2.35
Molecular weight	- 164.1
Melting point	- 555.7°C
Water Solubility at 100° C	-365 g 100 ML^{-1}

Effect of Soil

Due to presence of Ca it is a very good fertilizer for acidic soil. It should not be used in the soils having excess of moisture. It is equivalent to 21% of $CaCO_3$.

Effect of Crops: It is useful for sugarcane cotton, wheat, Jowar etc., crops because these crops require excess of Ca for their proper growth. It can be used in all type of crops.

Time and method of application

Due to deliquescent it becomes sticky very soon absorbing moisture, so it becomes difficult in applying it in the field. Therefore, it is used mixing with ammonium nitrate. It is broadcasted in standing crop.

AMMONIUM FERTLIZER

AMMONIUM CHLORIDE *(*NH_4Cl*):* It is called Muriate of Ammonia or Sal ammonia. It is more acidic, compared to ammonium sulphate. It is acidic fertilizer.

Manufacture: In India, it is prepared by several processes.

In this process, first NaCl salt brine of 30% concentration is ammoniated then, CO_2 is set in flow from which ammonium carbonate is formed.

$2NH_3 + CO_2 + H_2O \rightarrow (NH_4)_2 CO_3$

Ammonium carbonate is then converted to ammonium bicarbonate

$(NH_4)_2 CO_3 + + CO_2 + H_2O \rightarrow 2NH_4HCO_3$

By adding brine, NaCl solution, ammonium chloride is formed

$NH_4HCO_3 + NaCl \rightarrow NH_4Cl + NaHCO_3$

$$NaCI \xrightarrow{\text{unreacted}} NaCl$$

The reaction product still contain considerable amount of untreated sodium chloride. The sodium bicarbonate is sparingly soluble in this mixture of salt while ammonium chloride is soluble. Hence; sodium bicarbonate is separated by filtration or centrifugation. It first then heated to yield sodium carbonate (sodash)

$NaHCO_3 \rightarrow NaCO_3 + CO_2 + H_2O$

In other process, the solution of NH4Cl is concentrated, cooled and crystallized. The catalysts are then dried in a rotary drier at 105°C to reduce the moisture content to about 0.25%. The product is packed in moisture proof bag.

Properties:

Colour	White
State	Crystalline
Relatively	1.53
Melting Point	340°C
Molecular Weight	53.5
Water solubility at 100°C	75.8 g 100 ML^{-1}
Ammonium N	25%
Cl (Besides NH^4Cl)	1.5%

Carbonate (In Na_2CO_3 form)	0.5%
Sulphate (In Na_2SO_4 form)	0.5%
Insoluble residue	0.1%

Effect of soil: Ammonium chloride produces more acidity in soil than ammonium sulphate. So it should be applied in neutral or alkaline soil. It is decalcification fertilizer because it reacts with soil $CaCO_3$ and $CaCl_2$ which is lost by leaching. Excess use of NH_4^+ Cl as shown below:

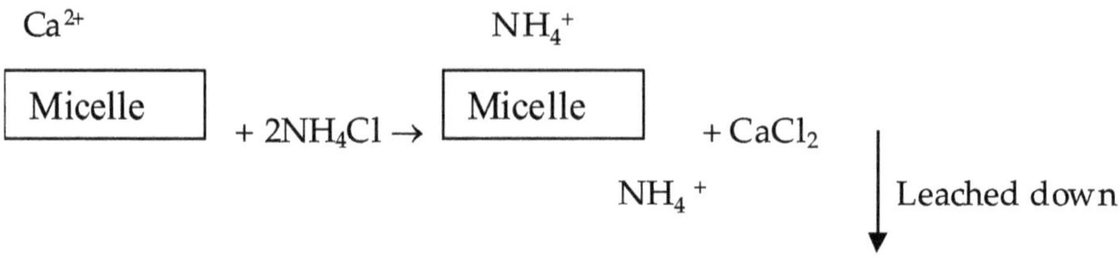

100 kg NH_4Cl produces acidity equivalent to 128 kg pure lime ($CaCO_3$)

Effect on crops: Except paddy crop it is taken up by other crop plants after nitrification. Excess use of NH_4Cl causes accumulation of Cl in plants which has toxic effect. $(NH_4)_2SO_4$ is better than NH_4Cl for most of the crops. The fertilizer value of NH_4CI is more than $(NH_4)_2SO_4$. Chlorine is also essential for plant in traces which is fulfilled by the fertilizer. Chlorine increases the yield of potato, tobacco, barley, cotton etc.

AMMONIUM SULPHATE $(NH_4)_2SO_4$: The most common nitrogenous fertilizer is ammonium sulphate. It contains 20.5% N and 24 % S. Its salt equivalent is 69 when $NaNO_3$ is standard (100). The pH of pure ammonium sulphate is 5.0.

Manufacture: In India, it is manufactured by two processes, namely (i) by product of coal and steel industries (ii) and by gypsum process.

(i) By product of coal and steel industries: Coal contain, on an average, about 1% N when coal is distilled in closed retarts in absence of air, its N evaporated as ammonia and ammonium salts which are dissolved in 'wash water'. This wash water is called ammonium liquor. The ammonia liquor is distilled with slaked lime from which ammonium gas is liberated. This ammonia gas is absorbed in H_2SO_4 to prepare $(NH_4)_2SO_4$.

$$\text{Coal} \xrightarrow[\text{Breaking}]{\text{Distillation}} NH_3 \xrightarrow{+H_2O} (NH_4)_2CO_3$$

$$(NH_4)_2\,CO_3 + CaCO_3 \rightarrow Ca\,CO_3 + 2NH_3 + 2H_2O$$

$$2NH_3 + H_2SO_4 \rightarrow (NH_4)_2SO_4)$$

Ammonia liquor

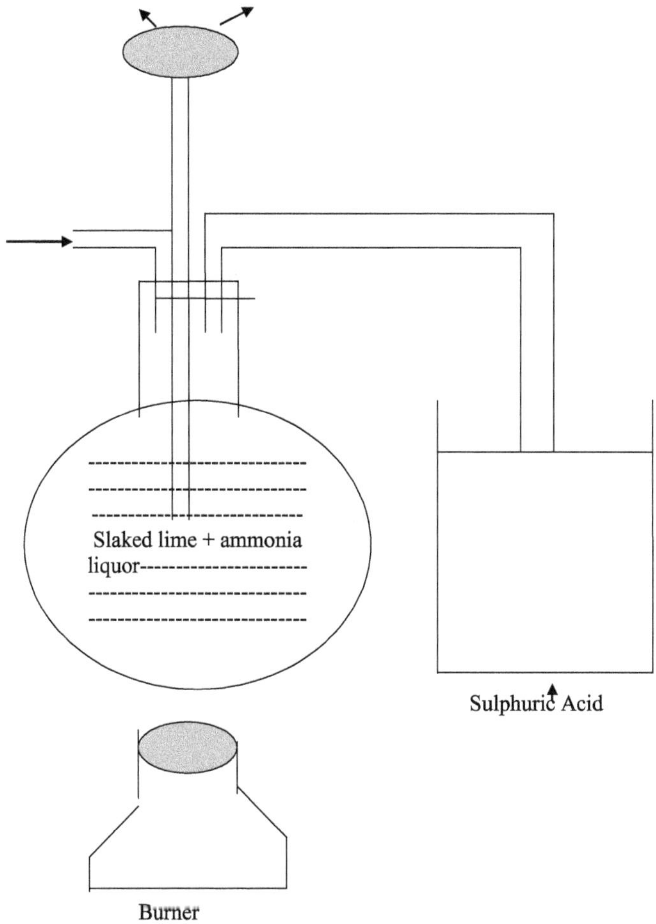

Fig.1. Preparation of ammonium sulphate

2.5-5.0 kg NH_3 is obtained from one ton coal from which 8-10 kg $(NH_4)_2SO_4$ is prepared by this method in Orissa, WB, MP, and Bihar. 14-15 kg ammonium sulphate is prepared from one ton steel in these factories.

(ii) By gypsum process: This is also called MERSBERG PROCESS. The raw materials required in this method are calcium sulphate, ammonia water and carbon dioxide. First NH_3 is prepared by synthesizing N_2 and H_2 gases which in the presence of water react with CO_2 and forms ammonia carbonate. By adding fine $CaSO_4$ in ammonium carbonate solution, ammonium sulphate and calcium carbonate are prepared. $CaCO_3$ is filtered which is used in preparation of lime or cement. The $(NH_4)_2SO_4$ solution is evaporated to obtain fertilizer. The following reactions occur in this process.

$$N_2 + 3H_2 \rightarrow 2\ NH_3$$

$$2NH_3 + CO_2 + H_2O \rightarrow (NH_4)_2\ CO_3$$

$$(NH_4)_2\ CO_3 + CaSO_4 \rightarrow Ca\ CO_3 + (NH_4)_2SO_4$$

Huge amount of $(NH_4)_2SO_4$ is prepared by this method in Albai (Kerala) and Sindari (Bihar). NH_3 is prepared by coal. About 0.6 tonne coal and 1.7 tonne gypsum is required to prepare 1 tonne ammonium sulphate. In India about 10 lac tonne AS is prepared per year of which 12% is prepared by first method and 88% by second method.

Properties:

Colour	White
State	Crystalline
Relatively density	1.77
Melting Point	512.2°C
Molecular Weight	132.14
Water solubility at 100°C	103.8 g 100 ML-1
Nitrogen	20.5%
Moisture	1.0%
Free acidity	0.04%
Arsenic	0.01%

Effect on soil:

NH_4^+ ions replace Ca^{2+} ions from colloidal complex which forms $CaSO_4$. The formed $CaSO_4$ is leached down causing the deficiency of Ca in soil.

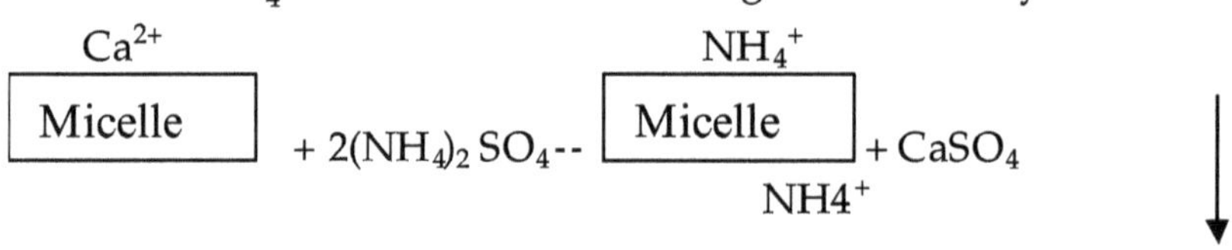

NO_3^- ions are produced due to nitrification of NH_4^+ ions absorbed on soil colloids which are taken up by plants or leached down.

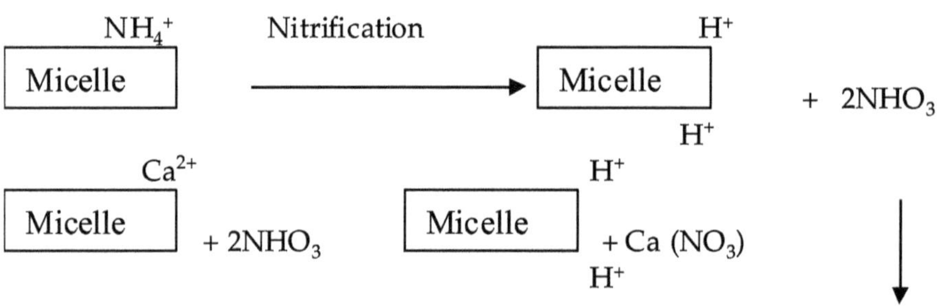

Thus, one $(NH_4)_2\ SO_4$ replace double of Ca from which great deficiency of Ca^{2+} ions on colloidal complex occurs and H^+ ion concentration increases causing acidity in soil. 110 kg $CaCO_3$ from soil is replaced by 100 kg $(NH_4)_2\ SO_4$ acidity in soil due to application of $(NH_4)_2\ SO_4$ is depend upon the following factors.

(i) Total Amount of ammonium sulphate

(ii) Duration and amount of rainfall

(iii) Amount of Ca^{2+} and Mg^{2+} present in soil.

Ammonium sulphate should not be added in acidic soil because it causes great loss of nutrient like P, K, and Ca. In soil having sufficient amount of lime, the $(NH_4)_2$ SO_4 can be used. $(NH_4)_2$ SO_4 can be applied in moist soil with standing crop because the NO_3^+ formed by nitrification of NH_4^+ions are being taken up by plants and loss of $(NH_4)_2$ SO_4 due to leaching does not occur. This can be applied successfully in both light and heavy soils.

Effect of crops

$(NH_4)_2$ SO_4 can be used in maize, potato barley etc crops. Small plants and paddy absorb N in ammoniacal form (NH_4^+). Besides paddy for which it is a good fertilizer, it is suitable for sugarcane, wheat, cotton and tea. This fertilizer also supply S to plants which is an essential plants nutrients thus, it is very good fertilizer for all type of crops.

Time and method of application

The $(NH_4)_2$ SO_4 can be used as basal and top dressing. It is also used as spray on crops. It should never be used with seed. In paddy crop before its use water should be removed from the field. It should only be applied at that place where irrigation facility is available. $(NH_4)_2$ SO_4 should be applied in two to three split in a crop. It should never be mixed with basic slag or lime because nitrogen is lost by becoming free ammonia-

$$(NH_4)_2\ SO_4 + CaO \rightarrow CasO_4 + H_2O + 2NH_3$$

Weedicide or fungicides having chlorate base should also be not used because there is danger of explosive.

ANHYDROUS AMMONIA (NH_3)

It is maximum concentration nitrogenous matter at normal atmospheric pressure and temperature being in gaseous form special type of instruments are needed for its storage, transport and use. Therefore, its use in India is very limited or it is not used.

Manufacturing and their process

Raw material: In synthesis of NH_3 nitrogen and hydrogen gases are needed. Atmosphere is the source of N in all the procedures. Hydrogen is prepared by petroleum product, natural gas and coal. In the production of ammonia raw materials used are divided into three groups.

(i) Solid- Coal, cock, and lignite

(ii) Liquid- Naphtha, liquefied petroleum gas and heavy fuel gas

(iii) Gas –Refinery natural gases and cock over gas

NH_3 production is completed in the following steps.

(i) Substances with C react with oxygen or vapour in presence of catalyst from which carbon monoxide and hydrogen is formed.

$CXH_4 + X\ H_2O \rightarrow ½\ YH_2 + CO$

(ii) Co reacts with vapour in presence of catalyst from which carbon dioxide and hydrogen is formed.

$$\underset{\text{(Excess of vapour)}}{CO_2 + H_2O} \xrightarrow[\text{iron and chromium}]{\text{Catalyst with}} CO + H_2O$$

(iii) In this step CO2 is removed by combining with mono hydroxyl ethylamine and pure hydrogen is remained

$2HOCH_2CH_2NH_2 + CO_2 + H_2O \rightarrow (HOCH_2CH_2NH_2)2CO_2 + H_2$

(iv) In presence of catalyst at very high pressure and temperature nitrogen and hydrogen combining together from ammonia.

$$N_2 + 3H_2 \xrightarrow[Fe_3O_4\ +\ Fe\ O]{\text{Catalyst}} 2\ NH_3$$

Properties:

Boiling point	33.03
Melting Point	77.0°C
Molecular Weight	17.03
Water solubility at 100°C	very high
Nitrogen	82 %
Moisture	1.0%
Ammonia	99%
Oil	20.0 ppm

Time and application

Anhydrous ammonia is added in dry soil 10-20 cm deep before one week of sowing, NH_4 ions formed by anhydrous ammonia with moisture, are absorbed on colloidal complexes which are taken up by plants. It produces acidity in soil. Its equivalent acidity is 148. It should be used ion those soils having excess use. It is not common in India due to unavailability of suitable and cheap instrument for its carriage.

Anhydrous ammonia is generally not explosive but when 16.25% anhydrous ammonia is present in air, it is explosive and inflammable.

Nitrate and ammonium fertilizers

AMMONIUM NITRATE: (NH_4NO_3)

In ammonium nitrate, half N is present ammoniacal form and half in nitrate form. Pure NH_4NO_3 contain 35% N but fertilizer contains 33.5% N.

Manufacture:

HNO_3 is prepared by setting in flow ammonia and oxygen gas on platinum catalyst. Ammonium nitrate solution is obtaining by neutralizing HNO_3 with NH_3.

$$NH_3 + 2O_2 \rightarrow HNO_3 + H_2O$$
$$HNO_3 + NH_3 \rightarrow NH_4NO_3$$

Properties:

Colour	White
State	Crystalline
Relatively density	1.73
Melting Point	170.4°C
Molecular Weight	132.14
Water solubility at 100°C	971.0 g 100 ML^{-1}
Nitrogen	33.5%
Ammoniacal N	16.75%
Nitrate N	16.75%
Equivalent Acidity	60.0 (Calcium carbonate)

It has self heating property due to that explosion may occur. There is a great problem in its storage. It should be stored in small heap in cold place.

Effect on soil:

It behaves as follow in soil.

$NH_4NO_3 \rightarrow NH_4$ + NO_3^- (↑ Used by plants; ↓ leaching)

NH_4 ↓ Nitrification → NO_3^- used by plants

It is acidic fertilizer. It is good fertilizer for alkaline and neutral soils. Its use is more beneficial in heavy soil than light soil due to less leaching in heavy soil.

Effect on Plants

It is best fertilizer for rain fed crops. It should not be used in paddy crop and in the area of excess rain fall because most of the nitrate is leached down.

Time and Method of application

Ammonium nitrate should be applied at the time of sowing as basal dressing or as top dressing in standing crop. It should be used less in Kharif season otherwise, great loss occurs due to leaching. It should be applied in split doses in light soil for chilli, cotton, sugarcane crops.

AMMONIUM SULPHATE NITRATE: $[(NH_4)_2SO_4.NH_4NO_3]$ ASN

It is a double salt. It is also called 'Sulpho nitrate or Leuna Salt Petre'. It is neither hygroscopic nor explosive.

Thus, this fertilizer is produced by mixing ammonium sulphate and ammonium nitrate.

Manufacture:

Ammoniation of the mixture of 50% HNO_3 + 77% H_2SO_4 is done for manufacturing $(NH_4)_2SO_4.NH_4\ NO_3$. The slurry is concentrated at 150°C to get 3% water remained in it. To check the formation of clod one tonne anti caking agent ($FeSO_4.7\ H_2O$) per 26 tonne nitrogen is added in the product. The product is cooled to 100°C and by chilling is converted into solid form. The excess acidity is neutralized by 10% NH_3 and granules are made by granulator. As a conditioner Kieslghur is coated on granules.

Properties:

Colour	White like dust
State	Crystalline
Molecular Weight	212.0
Moisture	1.0%
Water solubility at 100°C	971.0 g 100 ML^{-1}
Nitrogen	26.0%
Ammoniacal N	19.25% (75 % of Total N)
Nitrate N	6.75% (25 % of Total N)
Free Acidity (In HNO_3 form)	0.015%

Effect on soil

It is an acidic fertilizer of which equivalent acidity in 93. It is very good fertilizer for red and yellow soils of sea shore. It can successfully be used in alkaline and saline soils.

Effect on crops

It is a best fertilizer for all the crops. Ammoniacal-N releases slowly and nitrate becomes soon to small plants. Ammonium sulphate nitrate contains 12 % S besides N which is an essential plant nutrient. Thus, his fertilizer is very useful for the crops suffering from S deficiency. This is very useful for paddy.

Time and method of application

After applying ammonium sulphate nitrate in soil it should be at once mixed in the soil. It can be applied before sowing at the time of sowing or in standing crop by broadcasting. It should never be used with seed. This fertilizer should not be mixed with basic slag and lime because nitrogen is lost as ammonia gas.

CALCIUM AMMONIUM NITRATE ([Ca $(NO_3)_2NH_4NO_3$ CAN

This fertilizer is a combined form of two salts calcium carbonate and ammonium nitrate. Thus; it reduces the defects of ammonium nitrate fertilizer. It is also called 'KISHAN KHAD' or 'SONA' or 'NITRO CHALK': It is acidic in nature Application of 100 kg of this fertilizer makes the soil acidic. This acidity requires 93 kg of calcium carbonate to neutralize. Thus, AsN is less acidic compared to As.

Manufacturing: CAN is formed as follow:

$$\text{Air} \xrightarrow{\text{Condensation}} N_2 + \text{other gases}$$

$$4\ HOH \xrightarrow{\text{Electrolysis}} 2H_2 + O_2 + 2H_2O$$

$$3\ H_2 + N_2 \xrightarrow[\text{[Fe + Mo]}]{\text{High pressure and temperature}} 2NH_3$$

$$4NH_3 + 7O_2 \xrightarrow{\text{Pt}} 6H_2O + 4\ NO_2$$

$$2NO_2 + H_2O \longrightarrow HNO_3 + HNO_2$$

$$HNO_3 + NH_3 \longrightarrow NH_4\ NO_3$$

$$CaCO_3 + NH_4NO_3 \longrightarrow Ca\ (NO_3)_2.NH_4\ NO_3 + CO_3$$

NH_3 is prepared by Harber method. Some part of ammonia is oxidized to form HNO_3. Ammonium nitrate is formed with the combination of HNO_3 and remaining ammonia. Fine lime stone is added in concentrated ammonium nitrate to form CAN in granulation chamber. It is coated with soap stone to protect form moisture. It is prepared in Nangal fertilizer

Factory in India. It is slightly hygroscopic. During storage, it sets into cakes these cakes need crushing before use. It hygroscopicity is reduced by coating. It should be stored in dry and cool places.

Properties:

Colour	Brown or light brown
Molecular Weight	240.0
Moisture	1.0%
Water solubility at 100°C	very high
Nitrogen	20.0%
Total N	25% or 26.4%.
Ammoniacal N	12.5%
Nitrate N	12.5%
Calcium	8.1%

Effect on soil:

It is a neutral fertilizer. It does not causes acidity or alkalinity in soil so it can be applied in all types of soils. It can successfully be used in all types of soil specially in Ca deficient soil. It should be used in split doses in light. It is best fertilizer for heavy soils. It should never be used with basic slag and lime.

Effect on plants:

Calcium ammonium nitrate has at once effect on plants growth. It is very useful for sugarcane, cotton groundnut, potato, wheat etc. crops. Availability of P reduces due to its use. It can also be used in standing crops.

Amide fertilizers

CALCIUM CYNAMIDE ($CaCN_2$): It is also called nitro lime or Nitrilim. It is used as fertilizer and weedicide. It is not manufactured in India.

Manufacturing:

For its preparation lime stone and coal are required. The processes of manufacturing $CaCN_2$ in short as follows.

$$CaCl_2 \xrightarrow{\text{On heating in}} CaO + CO_2$$

$$3C + CaO \xrightarrow[\text{Electric furnace}]{} CaC_2 + CO$$

$$CaC_2 + N_2 \xrightarrow{1100°C} CaCN_2 + C$$

Properties:

Colour	Black
Molecular Weight	92.0
Nitrogen	20.0-21%
Total N	35% or 26.4%.
Free carbon	11%
Oil	50%
Moisture	2-3 %

It is a toxic substance. It produces irritation with skin.

Effect of soil:

It causes following reaction in soil –

$$2CaCN_2 + 2H_2O \longrightarrow Ca\ (HCN)_2 + Ca\ (OH)_2$$

Acidic soil

$$2Ca\ (HCN)_2 + 2H_2O \longrightarrow Ca\ (OH)_2\ CN_2 + 3H_2CN_2$$

Alkaline soil

$$2Ca\ (HCN)_2 + 2H_2O \longrightarrow (H_2CN_2)_2 + Ca\ (OH)$$

$$H_2CN_2 + 2\ H_2O \longrightarrow CO\ (NH_2)_2$$

Microbial activity

$$(CONH_2)_2 + 2H_2O \longrightarrow 2\ (NH4)\ 2CO_3$$

$(NH_4)_2CO_3$ converted into nitrate by nitrification and is taken up by plants. Some $CaCN_2$ is directly converted into urea. It is an alkali producing fertilizers. It should be used in acidic soil only.

Effect on plants/ crops: $CaCN_2$ has good effect on cotton, sugarcane, groundnut, paddy, wheat, jowar etc crops. It has toxic effect on plants. So, it should be used in grassland and posture.

Time and method of application:

It should be used 20-30days before sowing/transplanting. It should be mixed 8-10 cm deep in soil. It should never be used as top dressing or slide dressing.

Urea: [CO $(NH_2)_2$ or [NH_2CONH_2] or $CO \begin{matrix} \nearrow NH_2 \\ \searrow NH_2 \end{matrix}$

Urea has main place in nitrogenous fertilizers. In 1977 about 75% part of total N used for crops was by urea in our country. Now-a -days it increases to 80-85. Nitrogen present in urea is in amide form (NH_2). It is cheaper than other nitrogenous fertilizer @ one kg nitrogen. In India, Urea was first produced at Sindri (Bihar).

Manufacturing

First Scientist Ronelle (1733) separated urea from animal urine. Later on in 1828 Wholer synthesized urea by calcium cyanide. Urea is manufactured by reacting anhydrous ammonia and carbon dioxide gas under very high pressure in the presence of suitable catalyst. The reaction involved is represented in a simplified way as follows:

$$2NH_3 + CO_2 \longrightarrow NH_2CONH_2 \text{ (An unstable intermediary product)}$$

$$NH_2CONH_2 \longrightarrow CO\,(NH_2)_2 + H_2O$$

The urea solution is concentration to 99 % and is sprayed into a chamber where urea crystals are formed. In manufacturing urea following steps occur-

Hygroscopicity

It is hygroscopic. It absorbs moisture from the air when the relative humidity is 72% or higher at 30°C temperature. When the relative humidity of the air is lower than 72 % at 30°C temperature urea loses its moisture. Prilled urea is less hygroscopic than crystalline urea. Under alternate high and low humidity it cakes. If it is conditioned it does not cake; thus, there is no problem of storing and handling. However, once the bag is opened, the entire quantity should immediately be used.

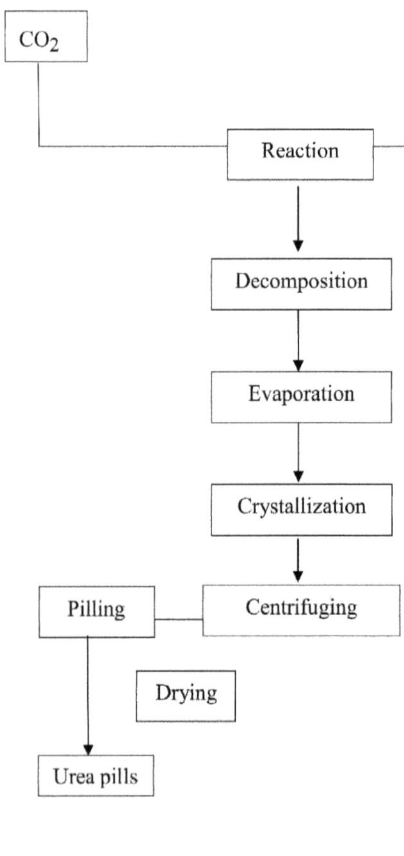

Properties:

Colour	White
State	Crystalline (Pure)
Granular (Fertilizer)	
Nature	Organic
Relative density	1.335
Melting point	132.7°C
Molecular Weight	60.06
Moisture	1.0%
Water solubility at 25°C	119 g 100 ML^{-1}
Total N	46.0%
Carbon	20.0%
Hydrogen	6.71%
Oxygen	26.64%
Biuret	1.5 %

Chemical formula

Its formula is $CO(NH_2)_2$. It is also called *carbamide* or *diamide* of carbonic acid. It is also called synthetic and non protein water soluble organic compound. It is identical with the urea found in animal urine.

Carbonic acid 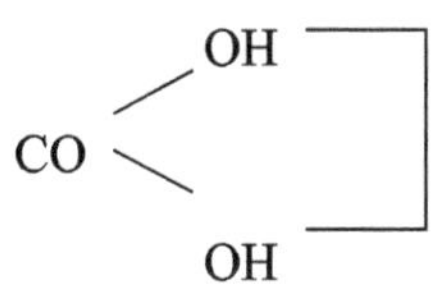

Here, OH is replaced by NH_2 to form urea.

Urea

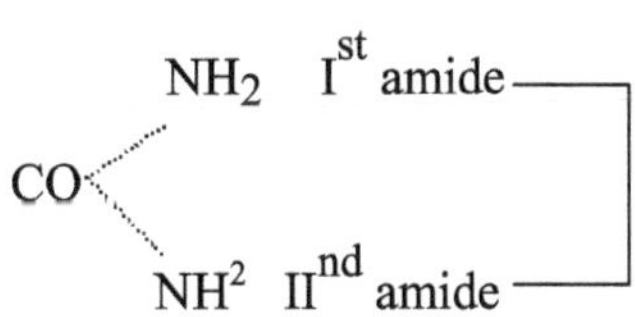

Thus, two amides of carbonic acid hence, diamide

Chemically pure urea contains 46.67 % N. The fertilizer grade urea contains 44% N. This amount is greater than amount found in any other N fertilizer.

Nature

It is an acidic fertilizer. Urea is less acidic compared to ammonium sulphate, since application of 10 kg of urea leaves acidity which requires 80 kg of calcium carbonate to neutralize.

Biuret content

When urea is heated to temperature above 170°C biuret is formed by elimination of one molecules of urea. During concentration and evaporation of urea solution to prepare either or crystalline prilled urea biuret is formed. Very little amount of biuret (less than 0.5%) is found in crystalline urea. The permissible maximum content of biuret in the Indian product is 1.5 %. Biuret is toxic to plant particularly when it is applied as foliar spray.

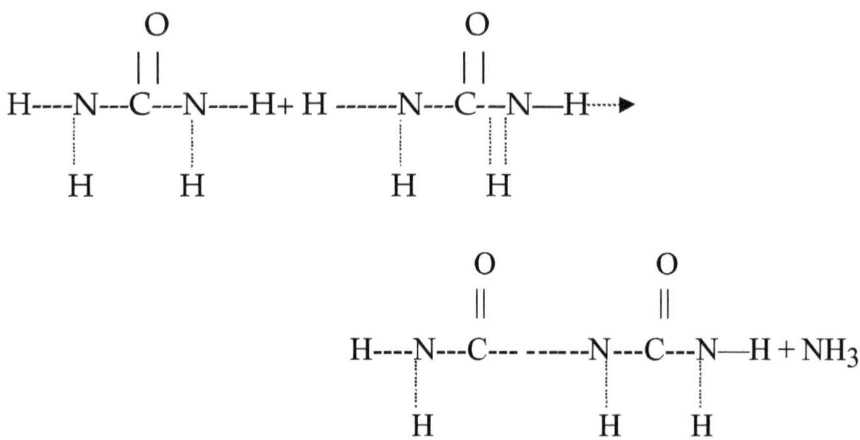

Biuret

Hydrolysis:

Urea is completely soluble in the soil solution but it is not utilized in amide form by most of the common field crops. Under field condition, it undergoes hydrolysis in the presence of urease enzyme by microorganisms and is converted into ammonium carbonate and then by microbial oxidation into ammonia and nitrate; and, in these form, is absorbed by plants. The probable changes urea undergoes after application to the soils.

$$CO\ (NH_2) + 2H_2O \longrightarrow (NH_4)_2\ CO_3$$

$$(NH_4)_2\ CO_3 + 3O_2 \longrightarrow 2HNO_2 + 3H_2O + CO_2$$

$$2HNO_2 + 2O_2 \longrightarrow 2HNO_3$$

Conversion of urea into ammoniacal and nitrate forms is complete in about a week. As such, it is advisable to supply urea three to four days before sowing of any field crop. Loss of nitrogen by leaching is negligible. The hydrolysis results in the loss of nitrogen by volatilization of ammonia. Hydrolysis of urea depends upon soil types, pH moisture and temperature. Hydrolysis occurs with more speed in fertile, clay neutral alkaline soil, in excess moisture and high temperature.

The obtained NH_4^+ ions undergoes as fallow:

- Nitrification occurs by nitrosomonas and nitrobacter resulting into NO_2^- and NO_3^- ions.
- Some part is directly taken up by plants.

- NH_4^+ ions are fixed on soil colloids.
- Some NH_4^+ ions are evaporated as NH_3.

Effect on soil:

The equivalent acidity of urea is 80. It should not be used in water logged soil. Most part of urea is taken up by plants as ammonical and nitrate nitrogen. Some amount of urea nitrogen is also absorbed by plants. Excess of urea in soil causes adverse effect on seed germination.

Effect on crops:

It is cheapest fertilizers. Urea can be applied in all type of crops at the time of sowing as basal dressing in standing crop as broadcasting and as foliar application also. It can also be applied as mixture with other fertilizer. It is equivalent useful to ASN and CAN and batter than AS. Its foliar application is very useful for most of the crops. It is very good fertilizer for sugarcane, groundnut jowar, bajra, wheat barley, potato oil seed crops fodder crop, vegetable etc. it should never be applied in deep water paddy crop. For dry farming it is one of the best fertilizers. Urea should always be applied in split doses to reduce the loss of N Foliar application of urea is more beneficial.

Advantages of Urea

i. Excess of Nitrogen percentage
ii. Easily and soon soluble
iii. Most cheapest fertilizer per kg nitrogen
iv. Foliar application is possible
v. Any residual effect doses not occur
vi. It can be used with other fertilizer and insecticides
vii. It is not explosive
viii. It can be used in soil as well as in standing crops
ix. It is less acidic than other nitrogenous fertilizers

Disadvantages of Urea

i. It is very soon hydrolyzed, due to that excess of nitrogen is lost as NH_3
ii. It is not suitable for fertilizer mixture
iii. Its denitrification occurs in excess.

Time and method of application

It can be applied before sowing, and at the time of sowing and in standing crops. It should never be used with seed because it causes scorching of seed. Due to high concentration it should be mixed with soil before use. It should be added in soil with super phosphate. It can be used as foliar spray. Solution of 2-3 % is generally sprayed. Absorption of urea by leaves in more than the absorption of urea by plant roots. Spray should be done after 30-40 days of sowing. Foliar spray can be repeated after 8-10 days. Some times excess of biuret is found in urea which causes great harm on spraying because biuret is a toxic substance. Biuret causes maximum loss to barley, wheat mulberry and orange. Urea can safely be used in soil if biuret is present more than 2%. The permissible maximum content of biuret in the Indian product is 1.5 %.

Slow available nitrogenous fertilizers: These are also called slow releasing fertilizers. By using slow releasing nitrogenous fertilizers, nitrogen loss is minimized. Generally, 50-60 % nitrogen of fertilizer is utilized by crops and the remaining N is lost by denitrification and leaching. For minimizing the loss of nitrogen, slow releasing nitrogen fertilizers have been developed. Slow/controlled release and stabilized synthetic fertilizers are grouped as follows:

Slow –release fertilizer	Coated fertilizers	Sulphur coated urea
		Polymer coated
	Uncoated fertilizer	Uncoated inorganic fertilizer
	Uncoated organic fertilizer	
	Super granules	
	Stabilized fertilizer	Nitrification inhibitor
	Urease inhibitor	

Main slow releasing fertilizers

1. Sulphur Coated Urea (SCU)- SCU was developed in 1961 by Tennesse Valley Authority, USA. In slow releasing fertilizers, SCU is maximum used Sulphur is good coating agent. It is cheap and essential plant nutrients. Sulphur is sprayed on urea pills on which wax containing 0.5% coal tar is coated. It contains 30-35% N which is slowly and slowly released in soil. It is also hygroscopic and is useful for those places where leaching occurs more it can successfully be used in paddy crop with better yield.

2. Urea form – By combining urea and formaldehyde in different ratio, various types of urea form is prepared. The availability of N depends upon the molecular ratio of formaldehyde and ammonia. Urea form is white and granular. It contains 38% N. A little amount of urea is also present in urea form. It is hygroscopic but caking does not occur. Nitrification of N occurs within 20-30 days but insoluble part is released slowly and become available to plant at the rate of 5-8% per month.

3. Isobutylidene diurea (IBDU) - It is prepared by the combination of urea and isobutyraldhyde in 2: 1 molecule ratio. It is white, crystalline non-hygroscopic and granular solid. It contains 30% N. It is very less soluble due to which it supplies N to plants slowly and slowly. At low pH in soil it is hydrolyzed into urea and isobutyraldehyde.

4. Oxamide –It is diamide of oxalic acid. Its production is very costly. It is white, crystalline non- hygroscopic solid. It contains 31.8% N. It is very less soluble in water. It is hydrolyzed in soil and slowly and slowly nitrification of NH_4^+ ion occurs into NO_3^- which is absorbed by plants.

5. Crotonylidene diurea (CDU) - It is the product obtained by the condensation of urea and acetaldehyde in the presence of acid catalyst. It contains 32% N. It solubility is 0.6g per litre at 20°C. It is useful fertilizer for wheat, sunflower and mustard.

6. N-Lignin- It is granular insoluble organic matter. It contains 18% N.

7. Guanal Urea- It is prepared in 3 form; sulphate, phosphate and nitrate salt. They contain 37.0, 27.8 and 37.0% N respectively.

Advantages

- Nutrient loss is less by leaching and run off.
- There is improvement in storage and transport properties.
- Less loss of N occurs by denitrification and volatilization.
- Quality of crop is improved.
- There is no any bad effect on seed germination.
- There is no need of top dressing in standing crops, thus, cost of top dressing is saved.
- Chemical and biological immobilization in soil is reduced.

Nitrogen in soil

The total nitrogen in soils generally varies from 0.02 to 0.04%.Due to tropical climate; the Indian soils are poor in total N which varies from 0.04 to 0.12 %.

Forms of N in soils

Nitrogen is present in different forms in soil. They are shown below:

Forms

- Organic
  - e.g. Protein, Acid, Free amino, Amino acid, Other complexes
- Inorganic
  - Ionic
    - e.g. NH_4^+, NO_3^-, NO_2^-, NO_2
  - Gaseous
    - e.g. N_2, N_2O, NO, NH_3

The form of nitrogen present in soil can be classed as organic and inorganic forms.

Organic Forms

The organic forms arise from plant and animal residues. They occur as protein free amino acid, amino sugar and other complexes. 20-40% of total soil nitrogen occurs as protein.

Inorganic Forms: The inorganic forms can be classed as ionic forms and gaseous forms. The ionic forms include ammonium ions, NH_4^+, nitrate ions NO_3^- and nitrite ions NO_2^-. 2-5% of total soil nitrogen occurs as these ionic forms. They arise from the mineralization of organic forms of nitrogen. Also, the fertilizers contribute these ionic forms. From standpoint of soil fertility they (except NO_2^-) are most important. The gaseous form include elemental nitrogen, N_2, nitrous oxide, N_2O, nitric oxide NO, nitrogen dioxide NO_2 and Ammonia NH_3.

Forms of Nitrogen absorbed by plants: Plants absorbed nitrogen in such forms as follows:

Nitrate ions, NO_3^-

They predominate in warm and well aerated soils. Hence, plants grown in well aerated soils absorb these ions.

Ammonium ions, NH_4^+

Urea, Co $(NH_2)_2$

Urea is absorbed through leaf epidermis, when urea solution is spread over the foliage. Usually it is not absorbed by roots.

Nitrate ions NO_2^-

Plants absorbs nitrate ions, NO_2^-. They are toxic. However, they accumulate in soil in small quantities.

Amino acid and nucleic acid

They are absorbed by higher plants. However, they accumulate in soil in small quantities.

Sources of Nitrogen in soil:

Soil gains nitrogen from fertilizers, organic manures, crop residues, biofertilizers, green manure organic wastes, rain water and bacterial nitrogen fixation.

Fertilizers and manures

The main sources of nitrogen for crop are fertilizers and manures which are applied to the soil during the course of crop rising. In these fertilizers, N is present either as NH_4^+ or NO_3^- or both or in amide (NH_2^-) form. Some of the ammonium –N- containing fertilizers are ammonium sulphate (20.6%N) Ammonium chloride (25%N) and DAP (18%N). Potassium nitrate (16% N) is a nitrate –N- containing fertilizers. CAN (25% N) ammonium nitrate (35% N) are the examples of fertilizers containing N in both forms. Urea, the amide form of nitrogenous fertilizers, containing 46% N is the most commonly used fertilizer in India.

Organic sources

These sources contain much less N as compared to the chemical fertilizers. Most commonly used organic sources are FYM (0.5-1.5% N), compost (0.5-2% N), oil cakes (3-8% N) etc. Organic sources are useful in maintaining the soil sustainability and are also eco-friendly. The organic sources contain less amounts of inorganic form of N (NH_4^+ and NO_3^-), but more amount of organic nitrogen.

Green manures

A 20 tonne weight of green manure can supply as much as 40-60 kg N ha^{-1}. Among the green manure crops, dhaincha (Sesbania *aculeate*) *Sesbania rostrata* and Sunnhemp (*Crotolaria juncea*) are commonly used as green leaf manure. For batter and quicker decomposition, 45-60 day –old green manure crop should be incorporated

Bio fertilizers

Among the bio fertilizers, Azolla and blue green algae are commonly used in rice culture. Azolla supplies about 30 kg N ha^{-1} whereas blue green algae provide 10-20 kg N ha^{-1} to rice crop (Singh, 1985). Besides, there are several bacterial fertilizers which are symbiotic *(Rhizobium)* and non- symbiotic *(Azotobacter)* fixers of nitrogen and are used in agriculture. These bacterial fertilizers fix atmospheric nitrogen to the extent of 10-20 kg N ha^{-1}.

Crop residues

Agricultural crop residues are also good sources of nitrogen. These materials are highly variable and may vary from 5 to 20 kg N ha^{-1}. Residues include roots of leguminous crops like green gram (*Vigna radiata*), black gram (*Phaseolus mungo*) and cowpea are also used by farmers as nitrogen sources. Depending on the biomass addition soil is enriched with nitrogen to the extent if 15-20 kg N ha^{-1} from these sources.

Organic wastes

Organic refuses of urban areas are composted through introduction of earthworm (wormi compost) or without earthworm and are used as a source of nutrients experiments on horticulture crops cereals and other crops have shown increases in crop yield due to the use of poultry dropping as it contains 1 to 3 % N.

Nitrogen from rain water

The nitrogenous compounds present in the atmosphere are brought down with rain water and snow to the earth's surface. These compounds are: NH_3, NO_2, NO_3, N_2O and organic N. On an average, 7 to 9 kg N ha^{-1} $year^{-1}$ is brought down by rain water and added to soil.

Biological Nitrogen Fixation: Biological nitrogen fixation and organic manures play a vital role in crop production. This is carried out by certain bacteria and actinomycetes, BGA, fungi algae and yeast. The most important of these are the bacteria. Bacteria have the ability to utilize the atmospheric nitrogen symbiotically in association with the plants or they may fix as free living organism (Non- symbiotically). However, symbiotic nitrogen fixation is of much importance and the amount of N fixed is quite appreciable.

Symbiotic Nitrogen fixation

Mechanism of fixation: Member of the genus *Rhizobium* bacteria infects the roots of appropriate legumes to develop nodules. The rhizobia live in the nodules. The legumes and rhizobia, the two dissimilar organisms, live together in intimate association (e.g., symbiosis). From the legumes, the rhizobia obtain their requirements for growth such as carbohydrates, mineral nutrients and other growth factors. The rhizobia take free nitrogen (e.g. Elemental nitrogen N_2) from the soil air and synthesize complex compounds of nitrogen (i.e., glutamic acid). Thus, nitrogen is fixed. The legumes absorb this nitrogen. The process of nitrogen is not well understood. It is assumed that the fixed nitrogen compounds in the cells of rhizobia are diffused through the cell wall. Thus, they (N compound) are absorbed by the legumes. The conducting tissue is formed between the main conducting tissue of the legumes and parenchyma which surround the colony of rhizobia. Thus, the transfer of substances between the legumes and the rhizobia is facilitated. Thus, symbiotic nitrogen fixation is the process by which elemental nitrogen N_2 is converted to its organic compounds by the association (or symbiosis) between two kinds of plants.

Fates of fixed nitrogen: The fixed nitrogen is subjected to several fates as follows.

- The fixed N may be used by the legumes
- When a legume and a non legume are grown in association, the fixed N (Aspartic acid, COOH-CH_2-CH (NH_2) - COOH) may be transferred from legume to non legume.
- When the legume sods are turned under some of the fixed N is quickly mineralized to ammonium, NH_4^+ and nitrate, NO_3^- ions that are absorbed by succeeding crops.

Amount of Nitrogen fixed

The amount of N fixed depends on several factors as follows:

- Soil properties
- Contents of nutrients
- Leguminous species

Thus, for a given legume the amount of fixed N varies from soil (i.e. Land). Nitrogen fixing capacity of the legumes may be graded as

Alfalfa> Soybean> Field bean> Field Pea,

In all lands (soil) alfalfa shows highest nitrogen fixing capacity while pea shows lowest. Legumes do not always increase the amount of soil nitrogen. Sometimes, they decrease the amount of soil nitrogen. It is noted that nitrogen fixation is maximum when the soil contains minimum inorganic nitrogen. The application of large amount of nitrogen fertilizers reduces the activity of Rhizobia and hence, reduces the extent of nitrogen fixation. However, a very small amount of nitrogen fertilizer may be applied so that the legumes seedling gets required (never excess) nitrogen until the Rhizobia establish on the roots.

Non-symbiotic Nitrogen fixation

Mechanism of fixation: Certain organisms are live independently in soil (e.g, Azotobacter) in water (BGA) and on leaves (e.g Beijerenkia). These free living organism absorb elemental nitrogen, N_2 from atmospheric and synthesize organic nitrogenous compounds. This phenomenon is referred to as non symbiotic or asymbiotic or free fixation of nitrogen.

The organisms live in rhizosphere or even at the surface of plant roots and intercellular spaces of epidermal roots cells. The C_4 plants (maize, sorghum and tropical forage) are the specific host plants for *Azospirillum lipoferum* while the C_3 plants (wheat, barley, oats, rye and rice).

Amount of Nitrogen fixed : The amount of nitrogen gained by soils non symbiotic nitrogen fixation varies widely (from few kilogram to about 50 kilogram per acre per year). It is to be noted that the N-fixating organism absorb inorganic N (NH_4^+ and NO_3^-) present in soil. Untill the nitrogen in soil be deficient the organism do not absorb atmospheric nitrogen. If the soil contains more organic matter, the organisms are supplied with more energy (needed for their life processes) and consequently, nitrogen fixation by them increases. The amount of N gained is very small. It varies from 1 to 60 kg per hectare per year.

Fixation by trees and shrubs

Like leguminous crops some leguminous trees cause nitrogen fixation. The trees develop nodules. The bacteria in these nodules fix free nitrogen by the symbiosis process. Mimosa and Acacia species are examples of such leguminous plants. In rice-based cropping system, *Glyricidia sepium, Leucaena leucocephala* and *Sesbania biospinosa* are used as green leaf manures.

Some non-legumes as those belonging to genus *Alnus (Betulaceae) Myric* and *Umptonia (Myricaceae), Casuarina* (Casuarinaceae) etc. have root nodules and fix N_2 the same way as do legumes.

Mineralization of organic nitrogen compounds: Mineralization of organic nitrogen can be defined as the process by which nitrogen in organic compound gets converted into the inorganic ammonium ions (NH_4^+) or nitrate ions (NO_3^-). The process of mineralization included several steps like amminization, ammonification and nitrification; these are discussed below.

Aminization: In this step, protein and proteinous compound breaks down to yield amines amino acids or amines, carbon dioxides, energy and other reaction products.

Protein $\rightarrow$ R $-NH_2$ + CO_2 + energy + other reaction products

This process is brought about by some heterotrophic soil micro organisms. They are mostly bacteria and fungi and possibly actinomycetes. Amino acids or amine formed due to aminization are either utilized by the microbes (immobilized) or mineralized further to NH_3 by the process of ammonification. However, this step (of process) is termed aminization.

Ammonification: It is defined as the process of transformation of amino to ammonia by microbial enzymes (Stevenson, 1975). These amino acids and amines are converted enzymetically by bacteria, fungi and actinomycetes to ammonia form.

$R\text{-}NH_2 + H\,.OH \rightarrow NH_3 + R\text{—}OH + \text{Energy}$

Two reactions may be suggested to demonstrate it (Reaction 1 & Reaction 2)

$$NH_3 + H_2O \longrightarrow NH_4OH \quad ....(1)$$

$$NH_4OH \longrightarrow NH_4^+ + OH^-$$

In this reaction ammonia NH_3 reacts with water H_2O to form ammonium hydroxide NH_4OH. Being unstable ammonium hydroxide NH_4OH breaks apart to yield ammonium ions NH_4^+ and hydroxyl ions OH^-.

$$2NH_3 + H_2CO_3 \longrightarrow (NH_4)_2 CO_3 \rightleftarrows 2 NH_4^+ + CO_3^- \quad ...(2)$$

$$NH_3 + HNO_3 \longrightarrow NH_4 NO_3 \rightleftarrows NH_4^+ + NO_3^-$$

In this reaction, ammonia, NH_3 reacts with carbonic acid, H_2CO_3, or nitric acid HNO_3 formed in soils to yield ammonium carbonate $(NH_4)2CO_3$ or ammonium nitrate, $NH_4 NO_3$ which in turn, breaks apart to yield ammonium, NH_4^+ and carbonate ions, CO_3^- or ammonium, NH_4^+ and nitrate, NO_3^-.

In both reactions the common products are ammonium, NH_4^+. Hence, this step is termed ammonification.

Nitrification: Nitrification is the process of enzymatic oxidation of ammonia to nitrate brought about by certain microorganisms. The ammonia ions are oxidized to nitrate by nitrifying bacteria (nitrobacter) in soil. It is a two step oxidation process, the conversion of NH_4^+ to NO_3^- and NO_2^- to NO_3^-. It can be shown below:

$$NH_4^+ \xrightarrow{[O]} HONH_2 \xrightarrow{-2H} \tfrac{1}{2}\ HONNOH \xrightarrow{[O]} NO_2^- + H^+ + \text{energy}$$

$$NO_2^- + \tfrac{1}{2} \xrightarrow{\text{Nitrobacter}} NO_3^- + 20\ \text{kcal}$$

The oxidation is carried out by the nitrifying bacteria. The first reaction is carried out by Nitrosomonas micrococcus Nitrospira and Nitroglea bacteria and the second one by Nitrobacter and Nitrosystis. The population of nitrobacter is usually from $1x10^5$ to $1x10^7$ per gram of soil. The process of nitrification results in the release of ionic hydrogen which causes acidity.

The nitrification rate in the soil where oxygen is not a limiting factor is quite high and is not a constraint. Nitrification is affected by several factors. These include supply of ammonium ions, C/N ratio and soil temperature. Depending on C/N ratio, variations of nitrification rate can be observed. At high C: N ratio, there would be immobilization of the ammonium ions and so the nitrification would be less. The reverse would be true in case of low C: N ratio. Small amounts of added nitrogen in such soils would stimulate nitrification.

Nitrification is not a problem in the neutral to alkaline soils (pH 7 to 9). The optimum pH is around 8.5. However, the range for nitrification is from pH 4.5 to 10.0. Even at pH values of 4.5, nitrates can be produced. Excessive NH_3 is toxic to nitrobacteria and inhibits the growth, especially of *Nitrobacter* and may results in high levels of NO_2^-, which is toxic to plants.

Factors affecting nitrification: The factors affecting nitrification are as follows:

- **Aeration** – Nitrification is an oxidation process. It needs oxygen. Thus, it takes place readily in well –aerated soils. Course textured soil (Sandy soils) granulated soils and ploughed soils ensure a good supply of oxygen. Hence, they promote nitrification.

- **Moisture-** In wet soils with very high moisture content nitrification is retarded because high moisture occupies the larger soil pores thereby reducing soil aeration (oxygen supply). In dried soils with very low moisture content also nitrification is impeded because nitrifying bacteria are killed in dried soils. The optimum moisture content for higher plants may be considered optimum for nitrification.
- **Temperature-** The ranges of temperature over which nitrification takes place in different soils are –
  - 5-35 °C
  - 0-2.0 °C
  - 16-30 °C
  - 27-32 °C

It indicates that the range of temperature in which nitrification take place varies from soil to soil. However, nitrification increases with increasing temperature, until the optimum temperature is reached. The optimum temperature varies with various soils. It may be 20-25 °C; 25-30 °C, and 30-35 °C, 35-40 °C. Nitrification does not take place at freezing points or below. It ceases at 51 °C or below.

- **Soil pH (Soil reaction)-**There is no clear relationship between soil pH and rate of nitrification. However, nitrification does not take place in strong acid soils, probably due to toxic effect of Al ions Al^{3+}, on nitrifiers. In some acid soils (pH 3.8 or 4.5) nitrifiers become adapted to acidity and thus, they cause nitrification. Here, the rate of nitrification is low. In some other acid soils with a pH above 5, nitrification take place only after the soils are limited. In strongly alkaline soils, the nitrite, NO_2, oxidizers (*Nitrobacter*) in the second step of nitrification, becomes sensitive to alkalinity. Thus, nitrate ions formation may be retarded or stopped. However, range of soil pH over which nitrification takes place is wide.
- **Lime-** The base such as Calcium, (Ca) and Magnesium, (Mg) increase the activity of nitrifiers. Hence, they enhance nitrification. Thus, to increase nitrification liming is recommended. The acid organic soils (having pH 5.5 or below) such as forest and peat soils contains high amount of lime (as high as 4% CaO). So, in these soils nitrification takes place.
- **Fertilizers-** If large quantities of urea or ammonia sulphate or other ammonium fertilizers are added to neutral or alkaline or calcareous soils free ammonia is formed. This free ammonia is much toxic to nitrobacter (Nitrite oxidizer). Hence; nitrate formation is retarded or stopped. However, free ammonia is less toxic to *Nitrosomonas* (ammonium oxidizers). Hence, nitrites, NO_2^- are formed considerably.

They accumulate in soil because their oxidation to nitrate, NO_3^-, is retarded or stopped Some times; the accumulated nitrites NO_2^- become *toxic.*

- **Carbon: Nitrogen Ratio (C: N Ratio)-** The ratio of carbon content (% C) and nitrogen content (% N) of substances is termed *carbon to nitrogen ratio.* It is expressed as C: N ratio or C/N ratio. Thus, C:N ratio

$$= \frac{\%C}{\%N}$$

The C: N ratio expresses the relative amounts of these two elements, and nitrogen. If the C: N ratio is wide the amount of carbon is much higher over the amount of nitrogen. If the C: N ratio is narrow, the amount of carbon is slightly higher over the amount of nitrogen. Note that the C: N ratio of any organic substances is not below 1. This C: N ratio influence the nitrification process. Suppose that large quantities of organic matters having a wide C: N ratio such as rice straw (C: N ratio 89: 1 or simply 89) wheat straw (C: N ratio 124:1 or 124) maize stalks and leaves (C: N ratio 53: 1 or 53) jowar stalk and leaves C: N ratio 93: 1 or 93) etc are incorporated in soil. As the C: N ratio is wide, the organic matters contains large amount of carbon or carbonaceous matter as compared to nitrogen. This carbon or carbonaceous matter supplies huge amount of energy to the heterotrophic microorganism, which decompose organic matter. Consequently, the microorganisms multiply rapidly. For multiplication, they need large amount of inorganic nitrogen (NH_4^+ and NO_3^-). Thus, they assimilate these inorganic nitrogen. During assimilation, the inorganic nitrogen is converted to organic nitrogen such as proteins, nucleic acid etc. The ammonium, NH_4^+ and nitrate, NO_3^- ions present in soil solution are mobile. Once, they are assimilated by the microorganisms they become immobile. Thus, this phenomenon is termed as immobilization.

When immobilization takes place, nitrification does not take place. Under this condition, very little or no nitrogen is available to plants. The plants may suffer from nitrogen deficiency. Thus, the nitrogen deficiency induced by immobilization is a temporary phenomenon. However, to prevent nitrogen deficiency, adequate amount of nitrogen fertilizers are to be added at the time of incorporated organic matters in soils such that the needs of both microorganisms and growing plants are met with. On an average, if organic matters having a narrow, C: N ratio (below 20: 1 or 20) are added to soils, mineralization take places. Usually immobilization does not take places. If organic matters having a wide C: N ratio (above 30: 1 or 30) are added to soils, immobilization take places during initial decomposition. *If organic matters having neither narrow, nor wide C: N ratio (between 20 and 30) neither immobilization nor mineralization take place.*

Losses of nitrogen from soils: Nitrogen is gained by soil through fertilizers manures; fixation by microorganism or few other sources but it does not accumulate in the soils or is used fully by the crops. Soil losses nitrogen from its stock through various processes during crop growth or fallow period which are discussed below.

- Denitrification
- Decomposition
- Volatilization Leaching
- Crop removal
- Erosion

Denitrification: The meaning of the words *nitrification* and *denitrification* are opposite to each other. Nitrification means Formation of nitrate, while denitrification means break down of nitrate. Indeed, nitrate is ultimately formed. The exact mechanism of this process is not known. Denitrification occurs in soils under conditions of poor drainage and poor aeration. It is caused by soil microorganism (facultative anaerobes). They are known as denitrifying organism or denitrifies. Under anaerobic condition of soils e.g. submerged or water logged and cloded or compact soils free oxygen, (O_2) present in nitrate (NO_3^-), nitrites (NO_2^-) and nitric oxide (NO) for their respiration. Thus, the nitrates, nitrites and nitric oxides are reduced to nitrous oxides and N_2O and ultimately to free nitrogen gas.

$$2NO_3^- \xrightarrow{-2O} 2NO_2^- \xrightarrow{-2O} 2NO \xrightarrow{-O} N_2O \xrightarrow{-O} N_2$$

Here, "–O" means removal of Oxygen (e.g. reduction)

Decomposition: While the reduction of nitrate (NO_3^-) is termed biological denitrification (shortly denitrification) the reduction of nitrite (NO_2^-) is termed non biological denitrification. This process is also termed non enzymatic decomposition or chemical decomposition or non biological decomposition or chemical denitrification or chemo denitrification or indirect denitrification

$$3HNO_3 \longrightarrow HNO_3 + H_2O + 2\,NO\uparrow \text{ Or}$$

$$2HNO_3 \longrightarrow NO_2 + H_2O + NO\uparrow$$

In these reaction, nitrous acid HNO_2 (e.g. nitrite) spontaneously decomposes in strongly acid soils (pH below 5) to yield nitric oxide, (NO) and nitrogen dioxides (NO_2), because it (HNO_2) is unstable in acid medium. In presence of oxygen (O_2) nitric oxide (NO), oxidizes to nitrogen dioxides (NO_2).

$$2NO + O_2 \longrightarrow 2NO_2$$

This nitrogen dioxide are formed directly from nitrite (HNO_2) react with water to yield nitrous and nitric acids

$$2NO_2 + H_2O \longrightarrow HNO_2 + HNO_3$$

Thus, no loss of nitrogen in the form of nitric oxides (NO) gas takes place in aerated soils. In soils where oxygen is completely eliminated (i.e. in anaerobic soils) nitrogen is lost in the form of nitric oxide (NO). However, the decomposition of nitrite (HNO_2) occurs rapidly at pH 4 or below and slowly at pH 6 to 7. It does not occur at pH 8 or above.

All the reaction shows that nitrogen is lost mostly in the form of elemental nitrogen and rarely in the forms of nitric and nitrous oxides, NO and N_2O, the chemical decomposition of nitrite occur in acid soils only. The various factors affecting the loss of nitrogen by chemical decomposition of nitrite as evident from the foregoing discussions are :

- Nitrite NO_2^-, accumulation in soil
- Soil reaction *i.e.* soil pH
- Soil aeration

Volatilization: In this process free ammonia is produced from ammonium containing fertilizers *viz.* ammonium sulphate $(NH_{4)2}$ $SO_{4,}$ ammonia nitrate, NH_4 NO_3 ammonium chloride, NH_4Cl etc. & ammonium forming fertilizers *viz.* urea, $CO(NH_2)$. This free ammonia escapes from soil to atmosphere. This process occurs in alkaline and calcareous soils. Even in acid soil too, this process occurs if large amount of urea is added.

Few examples to demonstrate the volatilization of ammonia are given below.

1. Aqua ammonia *i.e.* ammonium hydroxide (NH_4OH) is unstable in alkaline medium. Hence, if added to the surface of alkaline soils ammonium hydroxide (NH_4OH) decomposition to yield ammonia gas

$$NH_4OH \longrightarrow NH_3 + H_2O$$

2. When added to moist alkaline soils ammonia fertilizers dissociate to yield ammonium ions.

$$(NH_4)_2\ SO_4 \longrightarrow 2\ NH_{4+} + SO_4^{2-}$$
$$NH_4\ NO_3 \longrightarrow NH_4^+ + NO_3^-$$
$$NH_4Cl \longrightarrow NH_4^+ + Cl^-$$

The ammonium ions are converted to ammonia that volatilizes

3. When added to the surface of alkaline soil anhydrous ammonia volatilizes.

$$NH_3\ (Liquid) \longrightarrow NH_3\ (gas)$$

4. If ammonium fertilizers are added to the surface of calcareous soils (containing more than 10% free $CaCO_3$) ammonia volatilization occurs.

$$(NH_4)\ _2\ SO_4 + CaCO_3 \longrightarrow Ca\ SO_4 + (NH_4)_2\ CO_3$$

5. Ammonia volatilization from urea occurs in acid neutral and alkaline or calcareous soils.

$$CO\ (NH_2)_2 + H_2O \longrightarrow (NH_4)_2\ CO_3$$
$$(NH_4)_2\ CO_3 \longrightarrow NH_3 + CO_2 + H_2O$$

Factors influencing volatilization

- Cation Exchange Capacity (CEC)
- Soil pH
- Calcium Carbonate
- Temperature
- Dryness
- Fertilizers
- Fertilizer application method

Leaching: Removal of inorganic or mineral nitrogen from surface soil by percolation of water is called leaching. It is noteworthy that maximum leaching occurs in the form of nitrate (NO_3^- more than 99% leaching). Minimum leaching occurs in the forms of ammonium (NH_4^+ less than 1% leaching). Because ammonium added through fertilizers or formed through during mineralization is soon converted to nitrate which, in turn, transforms to nitrate so nitrate accumulates in soil. Due to repulsion between negatively charged nitrate ions and negatively charged soil particles, the nitrate ions are not adsorbed by soil particles. Thus, nitrate ions exists in soluble form in water, so with percolating water, nitrate ions leach down largely. Like nitrate, nitrite also exists in soluble form. Since its presence very small, its leaching occurs in traces. Due to rapid conversion of ammonium to nitrate ammonium accumulates in small amount. Moreover, ammonium ions are absorbed by soil particles by electrostatic forces of attraction between positively charged ammonium ions and negatively charged soil particles.

Leaching loss can be reduced by increasing water holding capacity through organic matter addition, split application of fertilizers controlled irrigation and using nitrification inhibitors like N-serve, Nimin and coated fertilizers.

Crop removal: Nitrogen removed by a crop can not be considered as its loss in real sense as the absorbed nitrogen is converted to protein in plants which is utilized by mankind and animals as food and fodder. Nevertheless soil loses that much of nitrogen from the system. For example, removal of nitrogen by one tonne of rice grain from unfertilized plot varied from 8.9 to 38.4 kg N ha^{-1}, while in balanced fertilized plots ranged from 14.0 to 44.5 kg N ha^{-1} depending upon the soil type, climate and variety. Similarly, one tonne of wheat grain remove approximately 20 kg N ha^{-1}. The removal of nitrogen from soil by this process increase if

- The crop is high yielding.
- The crop is capable of absorbing high amount of N.
- The number of harvest per year increase.

Erosion: The great loss of nitrogen is brought about by erosion due to removal of surface soil which contains most of soil nitrogen. Loss of N by erosion is more serious than any other nutrients.

Fates of nitrogen fertilizers in Soils

Fates of Ammonium Sulphate soils $(NH_4)_2SO_4$)

When added to soil ammonium sulphate dissociates to yield ammonium and sulphate ions in soil solution

$$(NH_4)_2\ SO_4 \rightleftharpoons 2NH_4{}^+ + SO_4{}^-$$

These are subjected to several fates as follows:

- Plant uptake
- Immobilization
- Fixation
- Volatilization
- Nitrification

A portion of ammonium ions, $NH_4{}^+$ is assimilated by plants. A portion of them may be immobilized if large amount of straw, stalk or other similar easily decomposable organic matters are added to soils. A portion of ammonium ions may be fixed by some clay minerals and other portion may be volatilized if the fertilizer is added to the surface of alkaline or calcareous soils, and other portion can be converted to nitrites by nitrification.

The nitrates are subjected to several fates as follows:

- Plant uptake
- Leaching
- Denitrification

A portion of nitrates, $NO_3{}^-$ is absorbed by plants and some of them is leached and converted to nitric oxide, NO, nitrous oxide, N_2O and elemental nitrogen N_2 gases by denitrification if the soils are anaerobic high in OM, calcareous, neutral, and alkaline.

The sulphate ions, $SO_4{}^{2-}$ released into soils are subjected to several fates as follows:

- Plant uptake
- Reduction
- Formation of sulphate salts
- Adsorption
- Immobilization

When AS is added to soil, it influences the following soil properties:

- Soil acidity
- Soil structure

When added to soil AS acidifies the soil; decreases soil pH or either increases soil acidities or decreases soil alkalinity

Note:

- To neutralize the soil acidity produced by 100 kg of As, 110 kg of pure $CaCO_3$ is needed.
- As Ca $(NO_3)_2$ is partly leached the $CaCO_3$ will be below 110 kg for 100 kg of As addition.
- As it contain impurities the amount of calcite or lime stone if used as a source of $CaCO_3$ will be above 110 kg for 100 kg of As addition.

The application of AS changes the soil pH, When applied only once, this fertilizer may not change the soil pH. The continued application for several year (say, 15 year) may decrease the pH of acid or neutral soil by nearly 1 unit (say, pH changes from 5.4 to 4.5 or from 6.7 to 6.0, while the fertilizer may not decrease remarkable the pH of calcareous soils. The effect of As, on soil pH is greatly influenced by several factors such as frequency and dose of fertilizer application degree of leaching soil Ca, Ca content, plant uptake of ammonium, NH_4^+ or nitrate, NO_3^-, soil buffering capacity etc.

Fates of Ammonium Nitrate in soils (NH_4NO_3): When added to soils, Ammonium nitrate NH_4 NO_3 dissociate to yield ammonium ions NH_4 and nitrate ions NO_3^-

$$NH_4\ NO_3 \longrightarrow NH_4^+ + NO_3^-$$

During nitrification ammonium ions NH_4^+ dissociate d from ammonium nitrate, release hydrogen H^+. These H^+ ions, acidifies the soil.

Fates of Calcium ammonium nitrate in soils ($CaCO_3$. NH_4NO_3): The components of calcium ammonium nitrate (CAN) are calcium carbonate and ammonium nitrate. NH_4NO_3, acidifies the soil. The acidity produced by NH_4 NO_3 is neutralized by another component Ca CO_3 of the fertilizers. Thus, this fertilizers is termed neutral or nonacid forming fertilizer.

Fares of Ammonium Sulphate nitrate in soils ($(NH_4)_2SO_4$. NH_4NO_3): The ammonium sulphate nitrate has two components such as ammonium sulphate and ammonium nitrate when added to soil they dissociate

$$(NH_4)_2\ SO_4 \rightleftharpoons 2NH_4^{\ +} + SO_4^{2-}$$

$$NH_4\ NO_3 \rightleftharpoons NH_4^+ + NO_3^-$$

In these reaction, three ions are formed i.e. ammonium nitrate and sulphate. Both the component $(NH_4)_2$ SO_4 and NH_4 NO_3 acidifies the soil.

Fates of Ammonium Chloride (NH_4CI): When added to soils, ammonium chloride, NH_4CI breaks apart to yield ammonium and chloride ions, NH_4^+ and Cl.

$$NH_4Cl \rightleftharpoons NH_4^{\ +} + Cl$$

The chloride ions may be absorbed by plants and take up chlorides as anions Cl^-. They may be lost from the soil by the process of leaching and react with alkali metals like Ca, Mg etc to form their chlorides. As these chlorides are water soluble, they may be washed out in drainage water. Thus, the Ca and Mg may be lost from surface soil. Ammonium ions are converted to nitrate by nitrification.

Fates of urea in soils: ($CO(NH_2)_2$): When added to soils, urea undergoes hydrolysis and is converted into ammonium carbonate and then by microbial oxidation into ammonia and nitrate and these forms is absorbed by plants. The hydrolysis is caused by an enzyme urease. This enzyme is produced by a number of bacteria, fungi and actinomycetes. To determine the hydrolysis of urea the following reactions have been suggested.

$$CO(NH_2)_2 + 2H_2O \longrightarrow (NH_4)_2CO_3$$
$$(NH_4)CO_3 + 3O_2 \longrightarrow 2HNO_2 + CO_2 + 3H_2O$$
$$2HNO_2 + 3O_2 \longrightarrow 2HNO_3$$

Conversion of urea into urea ammonical and nitrate forms is complete in a week. As such, it is advisable to apply urea three to four days before sowing of any field crop. Loss of nitrogen by leaching is negligible.

Ammonia reacts with water to form ammonium hydroxide. It dissociates to yield ammonium ions NH_4^+ and hydroxyl ions, OH^-. Urea acidifies the soil if ammonia volatilization does not occur. Thus, urea decrease the soil pH. In other words, it either increase soil acidity or decrease soil alkalinity. The process of hydrolysis of urea to form ammonium carbonate takes about 3-7 days. Since the process is essentially a microbial enzymatic one soil having higher substrates (or organic matter) will take less time for completion of hydrolysis. Urea hydrolysis is also faster in high pH soils. The acidification causes a loss of calcium, Ca from surface soils. Thus, to compensate the loss of calcium Ca, for application of 100 kg of urea, $CO(NH_2)_2$, 80 kg of pure calcium carbonate $CaCO_3$, is to be applied.

Note:

- The amount of pure $CaCO_3$ needed to compensate the loss of Ca caused by 100 kg of urea is needed is called acid equivalent of urea.
- Thus, the acid equivalent value of urea is not constant for all soils under all conditions.

Fates of sodium nitrate in soil: When added to moist soils, sodium nitrate dissociates to yields sodium ions Na^+ and nitrate ions NO_3^-.

$$NaNO_3 \longrightarrow Na^+ + NO_3^-$$

Fate of Sodium

- **Sodium neutralizes soil acidity or alkalifies.** Being alkali metal sodium induces alkalinity in soil. Consequently, the soil acidity decreases or the soil become alkaline. Hence, this fertilizer ($NaNO_3$) is physiologically alkaline, though it is a neutral salt. It increases phosphorus availability.

- **Sodium reduces the leaching loss of calcium or magnesium in soil.**
- **Sodium induces poor physical condition of soils.** It deflocculates the soil colloids. Consequently, soils become impermeable. Poor aeration and poor drainage of water exits. Hence, this fertilizer is not suitable for application to heavy soils and soils under arid or semi arid region where sodium carbonate, Na_2CO_3 accumulation is high.
- Sodium increase potassium availability by cation exchange.
- Sodium supplements potassium as a plant nutrient.

Nitrification Inhibitors: Nitrification inhibitors are synthetic chemicals or non synthetic (*i.e.* naturally occurring) substances that temporarily (for several weeks to months) inhibit nitrite, NO_2^-, formation from ammonium, NH_4^+ or nitrate, NO_3^- formation from nitrite, NO_2^-. Most of the inhibitors inhibit activity of *Nitrosomonas* bacteria that oxidizes ammonium, NH_4^+ to nitrite, NO_2^-. Some inhibitors inhibit activity of *Nitrobacter* bacteria that oxidizes Nitrite to Nitrate.

Few of the nitrification inhibitors are as follows:

Synthetic inhibitors

- N-Serve or [2-chloro-6-(trichlomethyl) – pyridine] or Nitrapyrin
- Potassium azide
- Dicyandiamide(DCD)
- AM (2- amino-4-chloro-6-methyl pyridine)
- 5-ethoxy-3-trichloromethyl-1, 2, 4-thiadiazole (Terrazole)
- N-2, 6-dichlorophenyl succinamide (DCS)

Non synthetic (or naturally occurring) inhibitors

- Neem cake
- Mahua cake
- Karanj cake

Urease inhibitors

They inhibit soil urease activity to reduce the extent of urea hydrolysis. Consequently, ammonium, NH_4^+ and nitrate, NO_3^-, formations become slow. Some of them are as follows:

- Hydroxamic acid
- Acetohydroxamate
- Thiocarbamate
- Quinones

- Thiourea
- Methylurea

However, they are not highly promising.

General Recommendations Regarding Suitability of Nitrogenous Fertilizer for Different Soil Conditions and Crops: Based on the four aspects discussed above, the following general recommendations can be made.

1. For paddy, ammoniacal and ammonia forming fertilizers, namely As and, A. Chloride Urea, should be used. If these are not available ammonia nitrate fertilizers like AsN, A. Nitrate and Calcium nitrate should be used.
2. For wheat, nitrate and ammonia forming fertilizers namely CAN, and As are either superior to or at par with ammoniacal and amide source, namely A. chloride and urea.
3. As, Urea AsN and CAN are equally effective for maize. However, under high rain fall As was better than urea and CAN because nitrates and urea are highly prone to leaching under such conditions.
4. Oilseeds and pulses prefer As over other N Carriers due to the special role of sulphur in their nutrition.
5. For cotton, generally different sources of N have been found to be equally effective. In jute, NH_4 sources proved superior to NO_3 sources, possible due to greater N losses via leaching and denitrification from the later.
6. For horticulture crops, there is very little information on the specific preference for NH_4^+ or NO_3^- forms by these crops under Indian conditions.
7. For other field crops, all nitrogenous fertilizers are equally effective.
8. On acid soils or soils low in lime or calcium continued use of As, urea, A. Chloride and AsN should be avoided. If these fertilizers have to be used on acid soils, lime application is desirable.
9. A nitrate fertilizer, like sodium nitrate, is the best suited for top dressing and side dressing when growing crops need nitrogen immediately.
10. Since nitrate fertilizers are easily leached, they should not be applied in large quantities in light sandy soil or during the rainy season. In sandy soil, the entire recommended dose of nitrogen should be applied in two or three installments, and during the rainy season, use of sodium nitrate should be avoided.
11. In the winter or rabi season the nitrogenous fertilizers should be selected on the basis of cheapness per unite weight of nitrogen, as all

nitrogenous fertilizers are equally effective and loss of nitrogen due to leaching does not usually occur.

12. For improved fertilizer N use efficiency, apply nitrogenous fertilizer in split doses, and incorporate or deep place fertilizers N into soil.

Phosphatic Fertilizers

Phosphorus is called MASTER KEY of agriculture. It is an essential component of every plant cell and protoplasm. It has second place in plant life after nitrogen. Phosphorus fertilizers are being used in India from 17th century. Those inorganic substances which are used to supply phosphorus to plans are called phosphatic fertilizers. Bone meal was mainly used as phosphatic fertilizers before the discovery of rock phosphate. The plant nutrient content of all phosphatic fertilizers is expressed in terms of percentage of phosphorus pentaoxide (P_2O_5).

Chemistry of phosphorus: Phosphorus exits in soils in both inorganic and organic forms. When Phosphorus burns, it is converted into phosphorus pentaoxide (P_2O_5), which readily dissolve in water to give metaphosphoric acid.

$$4P + 5\ O_2 \longrightarrow 2\ P_2O_5$$

$$P_2O_5 + H_2O \longrightarrow 2HPO_3 \text{ (Metaphosphoric acid)}$$

In warm water, this changes into orthophosphoric acid

$$2HPO_3 + H_2O \longrightarrow H_3\ PO_4 \text{ (Orthophosphoric acid)}$$

Orthophosphoric acid, commonly known as phosphoric acid contains three replaceable hydrogen atoms. One two, or all the there of these atoms are replaced, giving rise to three salts as indicated.

$$\underset{\text{Meta Phosphoric Acid}}{HPO_3} \xrightarrow{+H_2O} \underset{\text{Ortho Phosphoric Acid}}{H_3PO_4} \longrightarrow \underset{\text{Waters soluble}}{H_2PO_4} \longrightarrow \underset{\text{Citrate soluble}}{{}^{-}HPO_4^{2-}} \longrightarrow \underset{\text{Insoluble}}{PO_4^{3-}}$$

The type of phosphate ions present in the soil solution depends on the soil pH. In soils having neutral to slightly alkaline pH, the HPO_4^- ions is the most common form. As the soil pH gets lowered and it becomes slightly to moderately acidic, both HPO_4^- ion and the $H_2PO_4^-$ ions prevail. At higher soil acidities $H_2PO_4^-$ ions tend to dominated

$$\underset{\text{(Strong acidic solution)}}{H_2PO_4\ ^-} \underset{H^+}{\overset{H^-}{\rightleftarrows}} H_2O + HPO_42\ ^- \underset{H^+}{\overset{H^-}{\rightleftarrows}} \underset{\text{(Strong alkaline acid)}}{H_2O + PO_43^-}$$

Classification of Phosphatic Fertilizers: The phosphatic fertilizers, on the basis of solubility, can be classified into three groups.

1.Water Soluble: In this group phosphorus is present in Monocalcium Phosphate [Ca($H_2PO_4)_2$] form. Such fertilizers are:

	P_2O_5 %
• Super phosphate single, Ca $(H_2PO_4)_2$. Ca SO_4	16-18
• Super phosphate triple, Ca $(H_2PO_4)_2$. Ca SO_4	46-48
• Mono ammonium phosphate, $NH_4H_2PO_4$	48.0
• Diammonium phosphate, $(NH_4)_2HPO_4$	46.0
• Ammoniated Super phosphate	18.0

2.Citrate Soluble: The fertilizers of this group are soluble in citric acid and insoluble in water. Phosphorus is present as PO_42^- ion combined with Ca as $CaHPO_4$ or Ca_2 $(HPO_4)_2$. The main fertilizers are:

	P_2O_5 %
• Dicalcium phosphate, $Ca_2H_2(PO_4)$ or $CaHPO_4$	32.0
• Basic slag, $(CaO)_5P_2O_5$+ SiO_2	15-25
• Calcium metaphosphate, $Ca(PO_4)_3$	60-64

3.Insoluble: Those fertilizers, which are not soluble in water and citric acid, are grouped into insoluble phosphatic fertilizers. P is present in Ca_3 (PO_4) or PO_4^{3-} form.

	P_2O_5 %
• Rock Phosphate, $Ca_3(PO_4).CaF_2$	20-40
• Bone meal (Raw)	20-25
• Bone meal (steamed)	22-30

4. Other Phosphatic fertilizers:

i. Thermophos
ii. Kotka phosphate
iii. Phosphobacterin
iv. Pelophos
v. Rhenania Phosphate
vi. Coronet Phosphate

Manufacturing process and general properties of phosphatic fertilizers:

Water soluble phosphatic fertilizers

Single Super phosphate

Single super phosphate is also termed as ordinary super phosphate, normal super phosphate, den super phosphate or super phosphate of lime. However, its simple name is super phosphate. This name, super phosphate, is applied to the fertilizer produced by reaction of sulphuric acid with rock phosphate. Equal amount of each are mixed in a den fitted with Rotary where following reaction occurs.

$$3Ca_3\ (PO_4).CaF_2 + 7\ H_2SO_4 + 3H_2O \longrightarrow 3Ca\ (H_2PO_4)_2.H_2O + 7CaSO_4 + 2HF$$

↓

Super phosphate

The mixture is left for complete cooling which becomes solid. It is cut into small particles by rotary cutters or grinded and filled in sacks.

Manufacturing process:

The single super phosphate is manufactured by two processes as follow:

- Batch process
- Continuous process

Batch process

In this process, required quantities of pulverized rock phosphate and sulphuric acid are mixed for 1-2 minutes in a container known as mixer. Then the mixed material is dropped into a den and kept for 3-36 hours (in some plant, kept for 15 minutes). Here, most of the reactions between rock phosphate and sulphuric acid occur within a few hours.

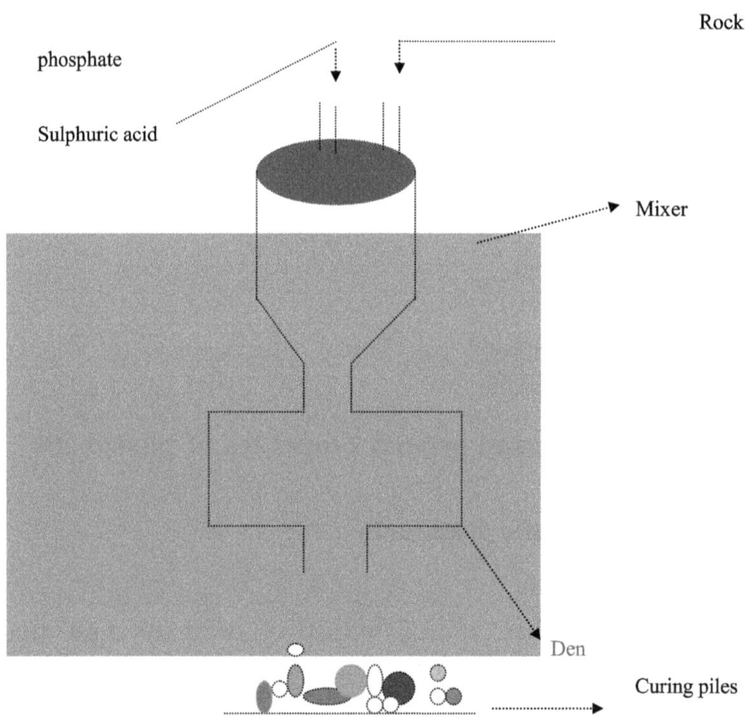

The water vapour, water carbon dioxide and hydrogen fluoride formed in the reactions escape from the product (curing) making it friable. It is then removed to curing piles where it is stored for several weeks (usually 2-10 weeks) here, the chemical reactions are completed, and the product is completed cured and dried. During curing, shrinkage (about 10%) occurs due to escape of water vapour, water CO_2 and HF. The product is then excavated, pulverized and bagged.

Continuous process: In this process, measured amount s of powdered rock phosphate and sulphuric acid are continuously mixed in a mixer for about 2-3 minutes. The mixture is then transferred to one end of the continuous den fitted over a slat conveyer. The mixture solidifies in the den. The solid product (SSP) is slowly moved to the other end of the den by the conveyer. Here the solid product is cut into small pieces by the cutter. The pieces are then transferred to curing piles for curing. The cured product is then bagged.

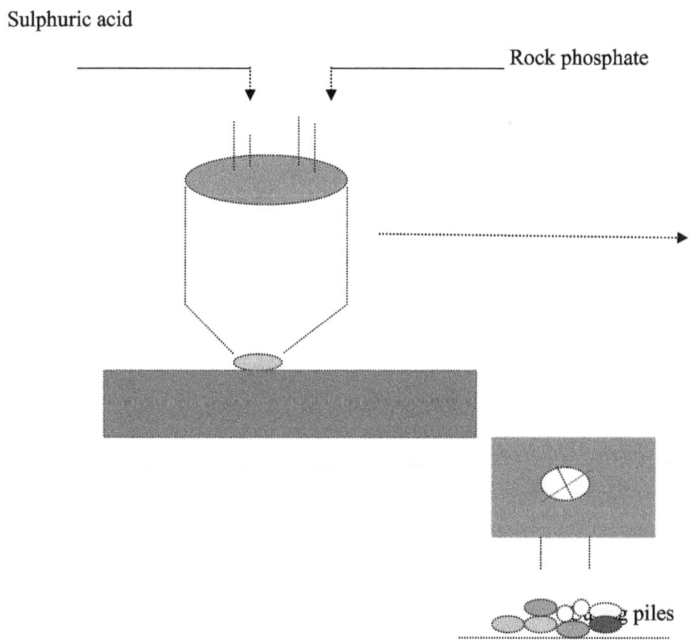

- It is unhygroscopic.
- It contains 10-22 % P_2O_5 of which about 85-89 % is water soluble.
- It also contains 10-12 % Sulphur.
- Colour is light brown.
- State is Non-crystal
- Water solubility is complete soluble
- Free phosphoric acid is (In the form of P_2O_5) -4%.
- Water soluble phosphate(In the form of P_2O_5)- 16 %
- Moisture is 12 %.

Effect on soil

On addition in soil, some part of super phosphate (mono calcium phosphate) is taken up by plants. In the meanwhile, within a very short period, the solution of super phosphate is precipitated in the soil pores. Depending on the soil pH, different phosphate compounds are formed. Thus, super phosphate is not lost by leaching.

In neutral to alkaline soils:

In neutral slightly acid and calcareous (alkaline) soils, dicalcium phosphate and tricalcium phosphate are formed as indicated

$$Ca(H_2PO_4)_2 + Ca(HCO_3)_2 \longrightarrow 2CaHPO_4 + 2H_2CO_3$$

Monocalcium Phosphate (Soluble); Calcium Bicarbonate; dicalcium Phosphate (Slowly available); Carbonic acid (lime)

$$Ca(H_2PO_4)_2 + 2Ca(HCO_3)_2 \longrightarrow Ca_3(PO_4)_2 + 4H_2CO_3$$

Tricalcium phosphate (Unavailable)

In acid soils:

In acid soils (pH below 7) iron and Aluminum react with super phosphate to form Fe and Al-phosphate which are unavailable to plant.

$$Al^{+3} + H_2PO_4^- + 2H_2O \longrightarrow 2H^+ + Al(OH)_2H_2PO_4$$

Aluminum ions; Dihydrogen Phosphate (soluble); Aluminum hydroxy phosphate (Insoluble)

$$2Ca(H_2PO_4) + Fe_2O_3 \longrightarrow 2FePO_4 + 2CaHPO_4 + H_2O$$

Insoluble and Unavailable

As phosphorus becomes unavailable in both acidic and alkaline conditions, measure precautions should be taken in its application:

1. Total P is basal dressed (10-12 cm below from surface) at the time of sowing.
2. It should be added with organic manures in acidic and alkaline soils.
3. It should be side dressed in long term crops i.e. sugarcane, cotton etc.
4. In very high acidic and alkaline soils, super phosphate should not be added.
5. It should not be top dressed in standing crops.
6. It should not be added in soil, with basic slag and lime in natural and alkaline soils.

Effect on Crops: Super phosphate single is batter than other phosphatic fertilizers because P present in it water soluble. Besides P it also supplies S to plants. It has better effect on leguminous crops and vegetables, berseem, pulse, groundnut potato, garlic, reddish, carrot get batter advantages from super phosphate. It has more effect on Rabi and Jayad crops than Kharif crops. It shows very good effect on tree plants. The quality of fruits is improved by the addition of super phosphate single.

Storage: During storage the water soluble monocalcium phosphate and citrate soluble dicalcium phosphate in super phosphate may partially be converted to insoluble tricalcium phosphate. However, the phosphorus in the newly formed tricalcium phosphate becomes slowly available to plants. If the super phosphate is stored a dry place the above mentioned conversion becomes nil or minimum.

If single super phosphate is mixed with ammonium sulphate and stored for a longtime ammonium sulphate reacts with monocalcium phosphate in super phosphate to form gypsum and monoammonium phosphate.

$$(NH_4)_2\ SO_4 + Ca\ (H_2PO_4)_2\ .H_2O \longrightarrow CaSO_4\ .2H_2O + 2\ NH_4\ H_2PO_4$$

Due to formation of gypsum, the mixture becomes hard which poses problems of application to soil. The addition of woody or fibrous organic matter reduces the hardness of the mixture. If single super phosphate is mixed with urea and stored for a long time, urea reacts with water soluble monocalcium phosphate in super phosphate to yield water insoluble dicalcium phosphate.

$$CO\ (NH_2)_2 + Ca\ (H_2PO_4).\ H_2O \longrightarrow CO\ (NH_2).\ H_3PO_4 + CaHPO_4 + H_2O$$

Under some conditions, the reaction between urea and monocalcium phosphate forms a complex. The above reaction shows the release of water from monocalcium phosphate monohydrate. It makes the mixture *sticky* and forms *cakes.*

If single super phosphate is mixed with calcium ammonium nitrate and stored for a long time, the water soluble monocalcium phosphate may be converted to water in soluble phosphate (reversion) due to the presence of lime in calcium ammonium nitrate.

Enriched Super phosphate: The term enriched super phosphate applies to a product obtained by the acidulation of rock phosphate with a mixture of sulphuric acids. It is essential that the rock phosphate is ground to about 75 % passing through a 200 mesh screen. Mahalingam (1974), has worked out the process for manufacture of enrich super phosphate containing 11-13% P (25.2-29.8 or 32% P_2O_5). The phosphoric Acid and sulphuric acids should be properly diluted and added to the conventional cone of single super phosphate plants. To reduce the power consumption in the mixer, it is preferable to add surface-active agents. The mixed super phosphate is sent from the mixer to conventional dens and then cured and bagged.

$$3Ca\ (PO_4)_2 + 6H_2SO_4 + 3H_2O \longrightarrow 3[Ca\ (H_2PO_4)_2H_2O] + 6CaSO_4$$

$$3Ca\ (PO_4)_2 + 12\ H_3PO_4 + 9\ H_2O \longrightarrow 9[Ca\ (H_2PO_4)_2\ H_2O]$$

$$3Ca\ (PO_4)_2 + 3H_3PO_4 + 18H_2O \longrightarrow 9[Ca\ (H_2PO_4)\ _2\ H_2O]$$

$$3Ca(PO_4)_2 \xrightarrow{\text{Unreacted}} 3Ca_3\ (PO_4)_2$$

$$3H_3PO_4 \xrightarrow{\text{Unreacted}} 3H_3\ PO_4$$

Enriched super phosphate

Modified Form of Single Super phosphate

In recent year, modification of common fertilizers have been developed and testedto give higher fertilizer use efficiency as compared to common fertilizers. Tandon (1991) has listed the following modification of single super phosphate.

I. **Boronated Single super phosphate:** SSP fortified with Borax (B) carries 0.18 % B in addition to 16 % P_2O_5. This material commercially available in the western country, particularly for the groundnut crop. It has been included in the Fertilizers Control Order.

II.* Zincated single Super phosphate:* ordinary super phosphate fortified with zinc sulphate to carry 205 % zinc has been developed and tested. The interest in Zincated SSP is largely in response to the deficiencies of both P and Zn widespread in Indian soils. The product is considered suitable for field use.

***III. Triple Super phosphate or Concentrated Super phosphate or multiple super phosphates*:**

Concentrated super phosphate is manufactured by processes which are similar to the processes for the manufactured of single super phosphate i.e. batch and continuous process.

Properties:

- **Composition:** The concentrated super phosphate is essentially monocalcium phosphate, Ca $(H_2PO_4)_2$. H_2O. Also it contain very little amounts of dicalcium phosphate, $CaHPO_4.H_2O$, tricalcium phosphate, $Ca_3\ (PO_4)_2$ and free phosphoric acid, H_3PO_4.
- **Odour:** the presence of free phosphoric acid gives the fertilizers an acidic odour.
- **Solubility:** This fertilizer is classified as water soluble fertilizers.
- **Chemical formula:** As this fertilizers is essentially mono calcium phosphate, its formula is Ca $(H_2PO_4)_2$. H_2O.
- **Content:** This fertilizer contains 19.9- 22.4% P (44-52% P_2O_5)
- **Other essential elements:** This fertilizer contains sulphur, usually 1.6% and calcium content is 14%.

- **Hygroscopicity:** this fertilizer is not disadvantageously hygroscopic.

Effect on soil and plant is similar to super phosphate single. The time and method of application is also same.

IV. Ammoniated super phosphate: Ammoniated super phosphate is produced by treating single super phosphate or concentrated super phosphate with ammonia reacts with components of super phosphate; the reactions are probably as follows:

$$2\ Ca\ (H_2PO_4)\ _2\ H_2O + 2NH_3 \longrightarrow 2NH_3H_2PO_4 + 2CaHPO_4 + H_2O$$

$$2CaSO_4 + 4NH_3 + 4\ CaHPO_4 \longrightarrow 2(NH_4)2SO_4 + 2Ca_3\ (PO_4)_2$$

$$H_3PO_4 + NH_3 \longrightarrow NH_4H_2PO_4$$

$$NH_4H_2PO_4 + NH_3 \longrightarrow (NH_4)_2HPO_4$$

Monocalcium phosphate, Ca $(H_2PO_4)_2$ H_2O, a component of super phosphate, reacts with ammonia to yield monoammonium phosphate, NH_4 H_2PO_4 and dicalcium phosphate, $CaHPO_4$. A portion of gypsum, $CaSO_4$ $2H_2O$ and dicalcium phosphate $CaHPO_4$ present in single super phosphate or formed in the previous reaction, react with ammonium, NH_3 to yield ammonium sulphate, $(NH_4)_2$ SO_4 and tricalcium phosphate, Ca_3 $(PO_4)_2$.Thus, the water soluble phosphate, Ca $(H_2PO_4)_2$ H_2O or Monocalcium phosphate in single super phosphate decreases and water insoluble phosphate, $CaHPO_4$ and Ca_3 $(PO_4)_2$ increase as the amount of ammonia added increase. Free phosphoric acid, H_3PO_4 reacts with ammonia to yield monoammonium phosphate, $NH_4H_2PO_4$. As added ammonia increase the monoammonium phosphate, $NH_4H_2PO_4$ formed in reactions is converted to diammonium phosphate, $(NH_4)_2HPO_4$.

Citrate Soluble Phosphatic Fertilizers

Basic Slag [(CaO) $5P_2O_5$. SiO_2]: It is also called Thomas Phosphate which is the by product of steel Industry. After super phosphate it is the second phosphatic fertilizers.

Manufacturing: Raw iron contains sufficient amount of P and traces of Ca, Si, and S. when steel is prepared from raw iron P is oxidizes into phosphorus pentaoxide P_2O_5 which combining with Ca Si and S impurities make dross which flows on melted metal. This dross is separated and made solid by cooling. Slug is prepared by grinding the dross. Presence of Ca makes it alkaline.

Properties

- It is hygroscopic, deep black brown with a very high specific weight colour fertilizers.
- According to international standard basic slag should contain minimum 13 % P_2O_5 of which 80 % part is soluble in citric acid.

- Basic slag also contains micro-nutrients like Cu, Zn, B, Mn which fulfils the requirement of plants for these nutrients.
- This fertilizer is especially suitable for acidic soils.

The most important point to be borne in mind, when using basic slag is that it should be mixed thoroughly with the soil because it hardly moves at all in soil.

Effect on soil: Phosphoric acid present in basic slag is easily soluble in soil water and become available to plants. CO_2 present in soil water increase its solubility. Mono-calcium phosphate is formed in soil as below which is available to plants easily.

$$(CaO)_5\ P_2O_5\ .SiO_2 + 6\ H_2O + 8CO_2 \longrightarrow Ca\ (H_2PO_4)_2 + 4CaH_2(CO_3)_2 + SiO_2$$

Basic slag contains Ca, Mg, Fe, and Mn which fulfills the deficiency of these nutrients in soil. It has better effect in acidic soil due to presence of excess lime. In acidic soil HPO_4^{2-} present in this combining with H^+ forms $H_2PO_4^-$ which is very easily available to plants. Thus, fixation of P is reduced.

Effect on Crops: P present in basic slag is easily and slowly available to plants due to that it is better for the crops like sugarcane, cotton, arhar etc.

It is one of the best phosphatic fertilizers for all those crops which grow in acidic and neutral soils. It is good fertilizer for pasture, grassland, paddy and groundnut.

Time and Method of application: Very fine power should be applied in soil. It should be applied one month before sowing because its reaction in soil is very slow. It should never be used with ammonium nitrate.

Dicalcium Phosphate ($CaHPO_4$) or [$Ca_2\ H_2\ (PO_4)$] : Dicalcium phosphate contains about 34 % of citrate soluble phosphate. This fertilizer has an excellent physical condition and due to its high P_2O_5 content is less costly for transport, storage and distribution. Since it contains citrate soluble phosphate, which will not turn into tricalcium phosphate, dicalcium phosphate or insoluble form as rapidly as monocalcium phosphate, dicalcium phosphate is suitable for wide range of soils and crops.

Manufacturing: It is prepared by the reaction of phosphoric acid on calcium oxide.

$$CaO+H_2PO_4 \longrightarrow CaHPO_4+H_2O$$

It is also manufactured by the action of phosphoric acid on rock phosphate.

$$[Ca_3\ (PO_4)_2]_3.\ CaF_2 + 4H_3PO_4 \longrightarrow 10Ca\ HPO_4 + HF$$

Properties

- It contains30-40%, P_2O_5.
- It is very less soluble in water but totally soluble in citric acid.
- Its storage and transport cost is very less because excess amount of P is present in dicalcium phosphate.

- It also contains 1.0 % Chloride 34% P_2O_5 and 8.0 % moisture.

Effect on soils: It can be applied in all type of soils but is more useful for acidic soil. In acidic soil P released by dicalcium phosphate is not easily fixed. It should never be used in calcareous soil because P fixation occurs very soon. Its equivalent alkalinity is 25.

Time and Method of application: It can be applied at any time in a year which should be basal dressed before one month of sowing. It can be kept after mixing with any fertilizer for a longer time. Therefore, it is generally used to prepare mixed and compound fertilizers.

Calcium Metaphosphate Ca $(PO_4)_2$: It is also called Metaphos.

Manufacturing: Phosphorus pentaoxide P_2O_5 vapour is obtained by burning P in a furnace. Rock phosphate is treated with P_2O_5 vapour to obtain calcium Meta phosphate;

$$[Ca_3(PO_4)_2]_3 \cdot CaF_2 + 7P_2O_5 + H_2O \longrightarrow 10\,Ca(PO_3)_2 + HF$$

Properties:

- It contains 60-64 %, P_2O_5
- Calcium oxide (CaO) is 26.5.
- Citrate soluble phosphorus is 53.0

Effect on soil: Calcium Meta phosphatic fertilizer is not useful for alluvial soil. It is a good fertilizer for acidic soil in which it can be used successfully.

Insoluble Fertilizers: Bone Meal $[Ca_3(PO_4)_2]_3 . CaF_2$: Bone-Meal is the oldest phosphatic fertilizer from ancient period. But bone meal is not very popular as phosphatic fertilizer in India. It is very popular as food material of hen and animals. Various types of bone meal fertilizers are produced and marketed for agriculture purpose. The following are the chief bone meal forms available in India.

Manufacturing: Bone meal is prepared by grinding bones. It is of two types:

1. Bone-Meal Raw or Untreated bone meal: Bones collected from city slaughterhouses and from the countryside are dried and crushed without treatments. Such raw bone meal contains 2-4 % nitrogen and 20-25 % P_2O_5.

2. Bone meal steamed or treated bone meal: The bones are sterilized with steam and then they are dried in warm, rotating ovens. After this process, the bones become brittle and can be fairly easily ground.

Properties:

- Total P_2O_5 in Raw bone meal is 25 % and 22 % in steamed bone meal.
- Available P_2O_5 in raw bone meal is 8 % and 16 % in steamed bone meal.

- Total N in raw bone meal is 3 % and 1.5 % in steamed bone meal.

Effect on Soil:

Bone meal is very useful in acidic soil having pH 3.5 -3.6. It is more useful in moist soil. Its effect remains in soil for many years. It has very less effect in neutral and alkaline soils.

Effect on Crops:

Bone meal is very beneficial fertilizer for annual crops as sugarcane, cotton, arhar, etc. It can be used in all types of crops. It has very good effect on tree plants because P present in this is being available for many years. N present in bone meal is in organic form which is very slowly available to plants.

Time and Method of application

Bone meal should be added in soil before one month of sowing. It should always be used with compost or FYM or green manuring. Bone meal should never be used in standing crops. It should be used with nitrogenous and potassic manures or fertilizers.

Rock Phosphate $Ca_3 (PO_4)_2$. CaF_2

There is huge stock of rock phosphate in India found in Trichanpalli (Tamilnadu), Udaipur and Jaisalmer (Rajasthan) Sirmaur (Himachal Pradesh) Purulia (West Bengal), Lalitpur and Mansoori (UP), Singh Bhoomi (Bihar), Jhabua (MP) and other places in India. It is a mineral which is generally used to prepare super phosphate single, triple and dicalcium phosphate etc. fertilizers. Now-a-days Indian rock phosphate is directly used as phosphatic fertilizer and has given good response especially in acidic soils.

Properties

- It is grey to dark in colour.
- It contains 15-35% P_2O_5 as insoluble tricalcium phosphate.
- It is called as Fluor Phosphate in market.
- It can directly be used in acidic soil having pH 3.5 to 6.5.

Quality of Rock phosphate: The rock phosphate needed for super phosphate production should have the following quality:

- The rock phosphate should contain at least 50 % Tricalcium phosphate, $Ca_3(PO_4)_2$
- It should not contain above 2 % iron oxide, Fe_2O_3. Reasons: If the iron oxide, Fe_2O_3 content is above 2 %, the consumption of H_2SO_4 for super phosphate production will be more. Also, the super phosphate

produced will be highly hygroscopic that pose problems of storing, Handling etc. high content of alumina, Al_2O_3 causes the same problem.

- It should not contain more than traces of pyrites, aluminosilicates, fluorides etc.

Reaction in Soil: When rock phosphate is added in highly acidic soil, insoluble tricalcium phosphate present in rock phosphate is converted into soluble and available forms. Excess of Fe^{3+}, Al^{3+} and H^+ ions are present in acidic soil which react with rock phosphate as below.

$$Al^{3+} + HOH \longrightarrow Al^{2+}(OH) + H^+$$

$$Fe^{3+} + HOH \longrightarrow Fe^{2+}(OH) + H^+$$

$$PO_4^{3+} + H^+ \longrightarrow HPO_4^{2+} + H^+ \longrightarrow H_2PO_4^-$$

Rock Soluble Phosphate (Insoluble and Unavailable) (Slowly available) Citrate Soluble (Easily available) Water

When rock phosphate is added in soil having excess of organic matter HNO_3 and H_2CO_3 present in soil convert tricalcium phosphate into soluble and available mono-calcium phosphate as shown below:

$$[Ca_3(PO_4)_2]_3.\ CaF_2 + 14\ HNO_3 \longrightarrow 3\ Ca(H_2PO_4)_2 + 7\ Ca(NO_3) + 2\ HF$$

$$[Ca_3(PO_4)_2]_3.\ CaF_2 + 6\ H_2CO_3 \longrightarrow 3\ Ca(H_2PO_4)_2 + 6\ Ca\ CO_3 + CaF_2$$

Insoluble and Unavailable Soluble and available

When rock phosphate is added in alkaline soil, it does not become available to plants. But when it is added in alkaline soil with inorganic and organic acidic matters as sulphur pyrites, compost FYM; green manuring etc becomes a little bit soluble and available to plants as shown below.

$$FeS_2 + 3O_2 \longrightarrow FeSO_4 + SO_2$$

$$2SO_2 + 2H_2O + O_2 \longrightarrow 2H_2SO_4$$

$$FeSO_4 + 2H_2O \longrightarrow Fe(OH)_2 + H_2SO_4$$

$$[Ca_3(PO_4)_2]3CaF_2 + 7H_2SO_4 + 3H_2O \longrightarrow 3Ca(H_2PO_4)_2 + 7CaSO_4 + 2HF$$

Rock phosphate (Insoluble and Unavailable) Mono calcium phosphate (Soluble and available)

Effect on Soil: When finely ground rock phosphate is applied to acidic soils or to soils containing a high percentage of organic matter, carbonic and nitric acid present in the soil act on rock phosphate, which contains tricalcium phosphate, and convert unavailable phosphate to monocalcium phosphate of water-soluble phosphate which is easily available to growing plants

$$[Ca_3(PO_4)_2]\ 3CaCO_3 + 6H_2CO_3 \longrightarrow 3Ca(H_2PO_4)_2 + 7CaCO_3$$

Rock phosphate Containing no flouring — Carbonic acid — Monocalcium phosphate — calcium carbonate

$$[Ca_3(PO_4)_2]3CaCO_3 + 14\ HNO_3 \longrightarrow 3Ca(H_2PO_4)_2 + 7Ca(NO_3)_2 + H_2CO_3$$

Rock phosphate — Nitric acid — Calcium nitrate — Carbonic acid

In general, ground rock phosphate should be applied in large quantities of 500 to 1000 kg per hectare once in five year but should not be applied in small quantities every year. It should be thoroughly mixed into the top soil by a country plough or disc.

Effect on Crops: Rock phosphate has batter effect on leguminous crops which need excess of calcium for their growth and development. It has less effect on short term crops and better effect on long term crops. It can be successfully be used in sugarcane, cotton and groundnut crops. It can also be used in other crops.

Time and Method of application: Very fine rock phosphate should be applied deep in soil before one month of sowing. It should be added in soil with compost, green manure, or FYM. It can be applied in soil with other fertilizers easily.

Phosphate Fixation: The fixation of phosphorus by soils has long been recognized. The fixation of P is a chemical reaction that makes it unavailable to crops. Another term often used in this sense is reversion. Soil pH has a deep effect upon phosphate fixation. In neutral to alkaline soils (pH 7 and above), the presence of lime (calcium carbonate) causes the precipitation of phosphorus. In acid soils (pH below 7) iron and aluminum react with phosphorus to form highly insoluble compounds. These reactions can be represented by the chemicals equations.

In neutral to alkaline soils:

In neutral slightly acid and calcareous (alkaline) soils, dicalcium phosphate and tricalcium phosphate are formed as indicated

$$Ca\,(H_2PO_4)_2 + Ca\,(HCO_3)_2 \longrightarrow 2CaHPO_4 + 2H_2CO_3$$

$Ca\,(H_2PO_4)_2$	$+\ Ca\,(HCO_3)_2$	$\longrightarrow$	$2CaHPO_4$	$+\ 2H_2CO_3$
Monocalcium Phosphate (Soluble)	Calcium Bicarbonate		dicalcium Phosphate (Slowly available)	Carbonic acid (lime)

$$Ca\,(H_2PO_4)_2 + 2Ca\,(HCO_3)_2 \longrightarrow Ca_3\,(PO_4)_2 + 4H_2CO_3$$

$Ca_3\,(PO_4)_2$: Tricalcium phosphate (Unavailable)

In acid soils:

In acid soils (pH below 7) iron and Aluminum react with super phosphate to form Fe and Al-phosphate which are unavailable to plant.

Al^{+3}	$+\ H_2PO_4^-$	$+\ 2H_2O$	$\longrightarrow$	$2H^+$	$+\ Al\,(OH)_2\,H_2PO_4$
Aluminum ions	Dihydrogen Phosphate (soluble)				Aluminum hydroxy phosphate (Insoluble)

$$2Ca\,(H_2PO_4) + Fe_2O_3 \longrightarrow 2FePO_4 + 2CaHPO_4 + H_2O$$

Insoluble and Unavailable

produced will be highly hygroscopic that pose problems of storing, Handling etc. high content of alumina, Al_2O_3 causes the same problem.

- It should not contain more than traces of pyrites, aluminosilicates, fluorides etc.

Reaction in Soil: When rock phosphate is added in highly acidic soil, insoluble tricalcium phosphate present in rock phosphate is converted into soluble and available forms. Excess of Fe^{3+}, Al^{3+} and H^+ ions are present in acidic soil which react with rock phosphate as below.

$$Al^{3+} + HOH \longrightarrow Al^{2+}(OH) + H^+$$
$$Fe^{3+} + HOH \longrightarrow Fe^{2+}(OH) + H^+$$
$$PO_4^{3+} + H^+ \longrightarrow HPO_4^{2+} + H^+ \longrightarrow H_2PO_4^-$$

Rock Phosphate (Insoluble and Unavailable) — Citrate Soluble (Slowly available) — Water Soluble (Easily available)

When rock phosphate is added in soil having excess of organic matter HNO_3 and H_2CO_3 present in soil convert tricalcium phosphate into soluble and available mono-calcium phosphate as shown below:

$$[Ca_3(PO_4)_2]_3.\ CaF_2 + 14\ HNO_3 \longrightarrow 3\ Ca(H_2PO_4)_2 + 7\ Ca(NO_3) + 2\ HF$$
$$[Ca_3(PO_4)_2]_3.\ CaF_2 + 6\ H_2CO_3 \longrightarrow 3\ Ca(H_2PO_4)_2 + 6\ Ca\ CO_3 + CaF_2$$

Insoluble and Unavailable — Soluble and available

When rock phosphate is added in alkaline soil, it does not become available to plants. But when it is added in alkaline soil with inorganic and organic acidic matters as sulphur pyrites, compost FYM; green manuring etc becomes a little bit soluble and available to plants as shown below.

$$FeS_2 + 3O_2 \longrightarrow FeSO_4 + SO_2$$
$$2SO_2 + 2H_2O + O_2 \longrightarrow 2H_2SO_4$$
$$FeSO_4 + 2H_2O \longrightarrow Fe(OH)_2 + H_2SO_4$$
$$[Ca_3(PO_4)_2]3CaF_2 + 7H_2SO_4 + 3H_2O \longrightarrow 3Ca(H_2PO_4)_2 + 7CaSO_4 + 2HF$$

Rock phosphate (Insoluble and Unavailable) — Mono calcium phosphate (Soluble and available)

Effect on Soil: When finely ground rock phosphate is applied to acidic soils or to soils containing a high percentage of organic matter, carbonic and nitric acid present in the soil act on rock phosphate, which contains tricalcium phosphate, and convert unavailable phosphate to monocalcium phosphate of water-soluble phosphate which is easily available to growing plants

$$[Ca_3(PO_4)_2]\ 3CaCO_3 + 6H_2CO_3 \longrightarrow 3Ca(H_2PO_4)_2 + 7CaCO_3$$

Rock phosphate Containing no flouring — Carbonic acid — Monocalcium phosphate — calcium carbonate

$$[Ca_3(PO_4)_2]3CaCO_3 + 14\ HNO_3 \longrightarrow 3Ca(H_2PO_4)_2 + 7Ca(NO_3)_2 + H_2CO_3$$

Rock phosphate — Nitric acid — Calcium nitrate — Carbonic acid

In general, ground rock phosphate should be applied in large quantities of 500 to 1000 kg per hectare once in five year but should not be applied in small quantities every year. It should be thoroughly mixed into the top soil by a country plough or disc.

Effect on Crops: Rock phosphate has batter effect on leguminous crops which need excess of calcium for their growth and development. It has less effect on short term crops and better effect on long term crops. It can be successfully be used in sugarcane, cotton and groundnut crops. It can also be used in other crops.

Time and Method of application: Very fine rock phosphate should be applied deep in soil before one month of sowing. It should be added in soil with compost, green manure, or FYM. It can be applied in soil with other fertilizers easily.

Phosphate Fixation: The fixation of phosphorus by soils has long been recognized. The fixation of P is a chemical reaction that makes it unavailable to crops. Another term often used in this sense is reversion. Soil pH has a deep effect upon phosphate fixation. In neutral to alkaline soils (pH 7 and above), the presence of lime (calcium carbonate) causes the precipitation of phosphorus. In acid soils (pH below 7) iron and aluminum react with phosphorus to form highly insoluble compounds. These reactions can be represented by the chemicals equations.

In neutral to alkaline soils:

In neutral slightly acid and calcareous (alkaline) soils, dicalcium phosphate and tricalcium phosphate are formed as indicated

$$\underset{\substack{\text{Monocalcium}\\\text{Phosphate}\\\text{(Soluble)}}}{Ca(H_2PO_4)_2} + \underset{\substack{\text{Calcium}\\\text{Bicarbonate}}}{Ca(HCO_3)_2} \longrightarrow \underset{\substack{\text{dicalcium}\\\text{Phosphate}\\\text{(Slowly available)}}}{2CaHPO_4} + \underset{\text{Carbonic acid (lime)}}{2H_2CO_3}$$

$$Ca(H_2PO_4)_2 + 2Ca(HCO_3)_2 \longrightarrow \underset{\substack{\text{Tricalcium phosphate}\\\text{(Unavailable)}}}{Ca_3(PO_4)_2} + 4H_2CO_3$$

In acid soils:

In acid soils (pH below 7) iron and Aluminum react with super phosphate to form Fe and Al-phosphate which are unavailable to plant.

$$\underset{\substack{\text{Aluminum}\\\text{ions}}}{Al^{+3}} + \underset{\substack{\text{Dihydrogen}\\\text{Phosphate (soluble)}}}{H_2PO_4^-} + 2H_2O \longrightarrow 2H^+ + \underset{\substack{\text{Aluminum hydroxy}\\\text{phosphate (Insoluble)}}}{Al(OH)_2H_2PO_4}$$

$$2Ca(H_2PO_4) + Fe_2O_3 \longrightarrow 2FePO_4 + \underset{\substack{\text{Insoluble and}\\\text{Unavailable}}}{2CaHPO_4} + H_2O$$

The reactions by which phosphate fixation takes place in soils have been grouped into three types:

1. adsorption
2. Isomorphous replacement and
3. double decomposition

The adsorption isotherms give an idea about the adsorption reactions. At low concentrations of phosphate in solution, the adsorption isotherm fits well either the Freundlich or the Langmuir equations. But at high concentrations and on prolonged contact other reactions take place.

The Isomorphous replacement of the hydroxyl ions with the phosphate ions is a possibility. The reactions of iron and aluminium hydroxides with the phosphate ions are probably most significant for the phosphate fixation in soils.

Factors affecting phosphate fixation: There are various factors which affect the phosphate fixation in soils like –

- nature and amount of soil components
- pH
- organic matter
- temperature, and
- Overliming.

1. **Nature and amount of soil components:** Adsorption and desorption reactions are affected by the types of surface contacted by P in the soil solution. However, various other soil components also affect phosphate fixation.
2. **Hydrous Oxides of Iron and Aluminum:** Soils with significant contents of amorphous iron and aluminium oxides have larger P fixing capacities than crystalline ones because of larger surface areas. Fixation of P by oxides of Fe and Al take place over a wide pH range as shown in below.
3. **Clay:** Capacity of soil to adsorb P is related to quantity and quality of clay. Soils containing large quantities of clay retain more P than those having small amounts of clay. Phosphorus is retained to a larger extent by 1:1 than 2:1 clays. There is more adsorption of phosphate in literate soils (Nilgiris) because of the presence of kaolinite as the dominant clay mineral and higher amounts of hydrous oxides of Fe and Al and the minimum phosphate adsorption is in black soils, being rich in smectite type of minerals which have practically no capacity for anion adsorption.
4. **Soluble Silica:** Capacity for P-sorption enhances with increased soil weathering (decreased Si solubility).The efficiency of utilization of P by rice has been found to increase with the addition of sodium silicate.

5. **Amorphous Colloids:** Amorphous aluminosilicates like allophane (Si-Al-Fe-O-OH-OH_2) have large negative charge which is party or entirely balanced by complex aluminium cations. P gets adsorbed by reacting with such aluminium ions.
6. **Calcium Carbonate**: In calcareous soils, calcite acts as the adsorbent site for P.
7. **Cations:** It is well known that more P is absorbed with increasing valency of the exchangeable cations in the system. Acid soils retain more P than alkaline ones whereas neutral soils adsorb more P than saline /sodic soils.
8. **Anions:** The effect of inorganic anions on P–sorption in alluvial laterite and black soils indicates that anions are more effective in reducing P-sorption capacity in coarse textured soil than fine textured soils.
9. **pH:** In strongly acidic soils, phosphate is fixed largely through precipitation with iron and aluminium in solution and in moderately acidic soil P is fixed by hydrous oxides of Al and Fe by anion exchange on silicates clays. In slightly acidic soils, P is fixed by hydrous oxides of Al and Fe by silicate clays. In alkaline soils, the soluble P from fertilizer reacts with Ca to form dicalcium phosphate which later becomes octacalcium phosphate but they do provide P to plants.

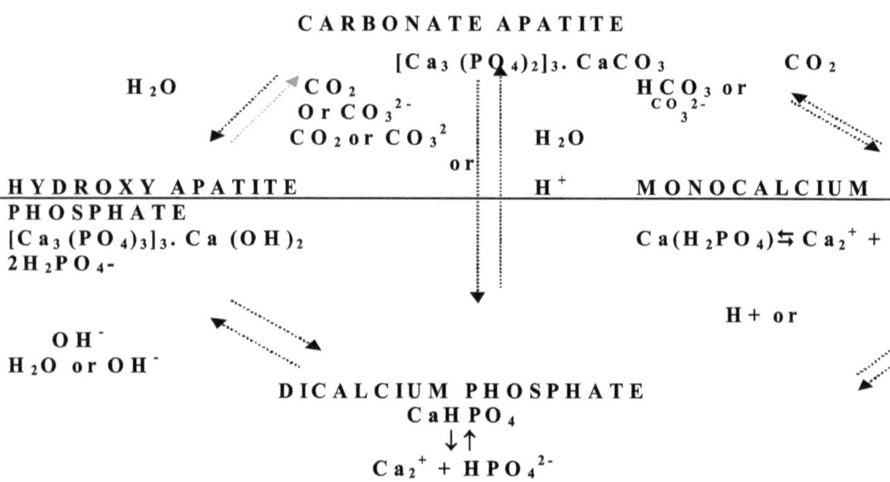

Fig. 2.1: Phosphate Cycle

Nitric (Nitro) phosphate; There are mainly three steps to produce nitric phosphate;

- *Decomposition* of apatite in rock phosphate by nitric acid to convert apatite to water soluble monocalcium phosphate. It is also known as acidulation or extraction.

- *Separation or* conversion of calcium nitrate
- *Ammoniation*

Some times, potassium salts e.g. KCl are added to the product to mark a complete fertilizer. To prevent the reversion of dicalcium phosphate, $CaHPO_4$, to tricalciumphosphate, $Ca_3 (PO_4)$, a stabilizer *e.g.* $MgSO_4$ may be added during manufacturing process. The product *i.e.* nitric phosphate may be pelleted or granulated and coated with anticaking material. Usually nitric phosphate is formulated for 30-50 % water solubility.

Properties: The component of nitric or nitro phosphate are as follows.

1. **Ammonium Nitrate:** Nitrogen is present in both ammonium and nitrate forms that are readily available to plants. The ratio of ammonium and nitrate may be 53:47 or 55:45.
2. **Monoammonium Phosphate:** it is soluble in water. Hence, its ammonium and phosphate are readily available to plants. The water soluble phosphate content is about 30% of total phosphate.
3. **Dicalcium Phosphate:** It is soluble in water. It soluble in neutral normal ammonium citrate (Citrate soluble). Thus, its phosphate is not immediately available to plants.
4. **Calcium Sulphate :** It is present if sulphuric acid or water soluble sulphate salt e.g.$(NH_4)_2SO_4$ is used in the manufacturing process of nitric or nitro phosphate provided the calcium sulphate is not removed
5. **Calcium Fluoride:** If the nitro phosphate is produced from rock phosphate containing fluorapatite, calcium fluoride is formed.
6. **NPK Content:** This fertilizer contains total N 6.4-8.6% total P 15-20 % P_2O_5 and total K is 0-12.4 %.
7. **Hygroscopicity:** It is slightly Hygroscopicity.

Ammonium Phosphate: It is also termed ammonium orthophosphates or ammophos –A

It is prepared by reacting phosphoric acid, H_3PO_4 with NH_3. The reactions that take place are as follows.

$$NH_3 + H_3PO_4 \longrightarrow NH_4H_2PO_4 \text{ (MAP)}$$

$$2NH_3 + H_3PO_4 \longrightarrow (NH_4)_2HPO_4 \text{ (DAP)}$$

Monoammonium phosphates, $NH_4H_2PO_4$ (MAP) or monoammonium orthophosphate and diammonium phosphate, $(NH_4)_2HPO_4$ (DAP) or diammonium orthophosphate are formed.

Properties

1. **Colour:** These fertilizers are light grey.
2. **Crystalline or granulated:** Usually, they are granulated.
3. **Hygroscopicity:** They are less Hygroscopicity.

4. **Acidic/neutral/basic fertilizer:** The MAP is acidic in reaction. DAP is slightly alkaline in reaction.

Urea Ammonium Phosphate: Wet process phosphoric acid is reacted with ammonia to yield MAP and DAP.

$$NH_3 + H_3PO_4 \longrightarrow NH_4H_2PO_4 \text{ (MAP)}$$
$$2NH_3 + H_3PO_4 \longrightarrow (NH_4)_2HPO_4 \text{ (DAP)}$$

Ammonium poly phosphate: Ammonium Poly phosphate is produced from super phosphoric acid.

Super phosphoric acid: When heated to high temperature (123°C) orthophosphoric acid H_3PO_4 undergoes dehydration *i.e.,* water molecules are eliminated from orthophosphoric acid molecules. Consequently, the molecules of orthophosphoric acid are condensed.

Manufacturing process: Super phosphoric acid is reacted with ammonia to yield ammonium poly phosphoric acid. In this reaction ammonia reacts each constituent phosphoric acid of super phosphoric acid *e.g.* orthophosphoric acid (*i.e.* monophosphoric acid), diphosphoric acids, tetra phosphoric acids and other higher phosphoric acids to form their ammonium salts.

$$H_3PO_4 + NH_3 \longrightarrow H_2PO_4\text{-}NH_4$$
$$2NH_3 + H_3PO_4 \longrightarrow (NH_4)_2HPO_4 \text{ (DAP)}$$

Properties

- Ammonium polyphosphate are concentrated or high analysis phosphatic fertilizers.
- They Contain 34-62% P_2O_5 and 10-15 % N.
- It is mixture of orthophosphates, pyrophosphate and other polyphosphates.
- It is highly soluble in water.

Phosphorus in soils

The form of P occurring in soils may be divided into broad categories, organic and inorganic.

Organic Forms: The organic forms of P originated from plant and animal residues and soil organism. They occur as –

- Inositol phosphate
- Phospholipids
- Nucleic acids
- Nucleotides
- Sugar phosphate
- Unknown forms

The organic forms comprises 30-50 % sometimes, more than 50% of total soil P. Inositol phosphate are most abundant. More than 50 % of total organic forms are present in unknown forms. Some of the organic phosphate exits in soil solution. Some organic phosphate (phytin, nucleic acid after its break down) may directly be absorbed by plants. The magnitude of this absorption is very low. The organic phosphates release inorganic phosphates by mineralization.

Inorganic Forms: The inorganic forms of phosphorus can be classed as

- Ionic forms
- Combined forms

Ionic forms: The ionic forms include primary orthophosphate, $H_2PO_4^-$ and secondary orthophosphate, HPO_4^-. These two ions are present in soil solution. Hence, they are termed solution phase phosphate. The both ions are equal at pH 7.00 – 7.22. In strongly alkaline solution, tertiary orthophosphate ions, PO_4^- dominate. As pH decrease the tertiary orthophosphate ions PO4- are converted to secondary orthophosphate ions HPO_4^-.

$$PO_4^{3-} + H^+ \longrightarrow HPO_4^{2-}$$

As pH decrease H+ concentration increase This H^+ react with PO_4^{3-} to yield HPO_4^{2-}. With further decrease in pH, the secondary orthophosphate ions HPO_4^{2-} are converted to primary orthophosphate ions $H_2\,PO_4^{2-}$.

$$HPO_4^{2-} + H^+ \longrightarrow H_2\,PO_4^{2-}$$

In strongly acid solution, primary orthophosphate ions, $H_2\,PO_4^{2-}$, dominate. As the pH of solution increases, the primary orthophosphate ions, $H_2PO_4^{2-}$ are converted to secondary orthophosphate ions, HPO_4^{2-}.

$$H_2\,PO_4^{2-} + OH^- \longrightarrow HPO_4^{2-} + H_2O$$

As pH increases the hydroxyl ion, OH^- concentration increases. These hydroxyl ions OH^- react with primary orthophosphate ions, HPO_4^{2-} to yield secondary HPO_4^{2-}. With further increase in pH, the secondary orthophosphate ions HPO_4^{2-} are converted to tertiary orthophosphate ions, PO_4^{3-}.

$$HPO_4^{2-} + OH^- \longrightarrow PO_4^{3} + H_2O$$

Thus, the concentration of ionic forms of P in soil solution is very low.

FORM OF PHOSPHORUS ABSORBED BY PLANTS

Plants absorb phosphorus in ionic forms as follows.

- Primary orthophosphate ion (or monovalent phosphate ion), $H_2PO_4^{2-}$.
- Secondary orthophosphate ions (divalent phosphate ion), HPO_4^{2-}.

The primary orthophosphate ion (or monovalent phosphate ion), $H_2PO_4^{2-}$ are absorbed more readily than secondary orthophosphate ions (divalent phosphate ion), HPO_4^{2-}. They are absorbed by plants from solution phase

Mineral	Formula	% K_2O
Silvite	KCl	63.1
Kenite	$4KCl.4MgSO_4.11H_2O$	19.2
Kornelite	$KCl\ MgCl_2.6H_2O$	17.0
Polyhallite	$K_2SO_4.MgSO_4.2CaSO_4.4H_2O$	15.6
Nitre	KNO_3	46.5
Leonite	$K_2SO_4.MgSO_4.4H_2O$	25.7
Langbeinite	$K_2SO_4.2MgSO_4$	22.7
Glosarite	$3K_2SO_4.Na_2SO_4$	41.5

Manufacturing and properties of important potassic fertilizers: The manufacturing process and properties of common used fertilizers are-

Potassium Chloride or Muriate of Potash: (KCl): It is most popular potassic fertilizer in the world. It is contains 60-96 % K_2O, and is guaranteed to 60% potash. It is manufactured from potash –bearing minerals. Two types of refining process are employed to separate the potassium chloride from other compounds present in mineral. There are two methods of producing KCl fertilizers.

i. Floatation

ii. Fractional Crystallization Method

Floatation: This process is based on the difference in the specific gravities of KCl and NaCl with KCl having lesser specific gravity floating on top of NaCl, which is separate and dried for use.

Fractional Crystallization Method: This method is totally based on the difference between the temperature solubility relations of KCl and NaCl. The fundamental principal involved in this process is that potassium chloride is much soluble in hot water than in cold, while the solubility of NaCl which is the main impurity does not vary much with a change in temperature.

Properties: Pure KCl is white and crystalline. Due to impurities the colour of KCl prepared by Floatation method is pink red. The main properties are as follows.

Colour	White
State	Crystalline
Molecular weight	74.56
Specific gravity	1.988
Water solubility at 100°C	56.7 g $100ML^{-1}$
Melting point	790.0 °C
% K_2	50
% K_2O	60-63
% NaCl	3.0
% moisture	0.5

Effect on soil: It is a neutral fertilizers and highly soluble in water. It is ionized in soil into K+ and Cl-. Some ions are fixed as follow-

$$KCl \rightarrow K^{+} + Cl^{-}$$

Fertilizer absorbed by plants

$$[\text{Micelle}]\,Ca^{2+}\ 2K^{+} \rightleftharpoons [\text{Micelle}]\ 2K + Ca^{2+}$$

$$[\text{Humus}]\ H^{+} + K^{+} \rightleftharpoons [\text{Humus}]\ + K + H^{+}$$

$$K^{+}\ \text{solution} \rightleftharpoons \text{Exchangeable}\ K^{+} \rightleftharpoons \text{Fixed}\ K^{+}$$

Easily and soon available to plants unavailable

Some K^+ ions are fixed in clay colloid complexes and some are absorbed on them. In acidic soil Cl^- ions take the place of OH^- ions related to free iron oxide. The effect of KCl in acidic soil is batter than K_2SO_4. In alkaline soils, Cl^- ions show toxic effect. So it should be added in alkaline soil with compost, FYM or green manuring.

Effect on plants or crops: Potassium chloride is the best potassic fertilizers for most of the crops. It can successfully be used in wheat, barley potato, leguminous and tuber crops. It should not be used in tobacco linseed and beet like crops because chlorine present in fertilizer deteriorates their quality.

Time and Methods application: It can be used before sowing at the time of sowing and in standing crop as side dressing or spray. Generally it is used at the time of sowing. Its foliar application has shown better effect on crop yield. It should be used with calcium nitrate.

Potassium Sulphate (K_2SO_4): It is also called sulphate of potash. In comparison to potassium Chloride, ***it is costly fertilizers***. It is the main non-chloride form of potash fertilizers. Pure potassium sulphate contains 54% K_2O and the commercial fertilizers contain 48-52 % K_2O.

It is commonly manufactured in the following two ways.

- **Langbeinite Mineral Method:** By dissolving langbeinite (Potash mineral $K_2SO_4.2MgSO_4$) in water and adding a concentration solution of potassium chloride. The potassium sulphate precipitates out and is separate by decantation.

$$K_2SO_4.\ 2MgSO_4 + 4KCl \longrightarrow 3\ K_2SO_4 + 2MgSO_4$$

Langbeinite potassium chloride Pot. Sulphate Magnesium Chloride

- By treating potassium chloride with concentrated sulphuric acid

$$2KCl + H_2\ SO_4 \longrightarrow K_2SO_4 + 2HCl$$

Potassium chloride sulphuric acid Pot. Sulphate hydrochloride acid

Properties: K_2SO_4 contain about 17 % SO_4 which is essential as S for plants. It is highly soluble in water. It is non- hygroscopic so it can easily be stored. It main properties are-

Colour	White
State	Crystalline
Molecular weight	174.56
Specific gravity	2.66
Water solubility at 100 °C	24.1 g $100ML^{-1}$
Melting point	1266.7 °C
% K	40-42.5
% K_2O	48-60

Effect on soil: It reacts in soil similar to KCl. It is very easily ionized in K^+ and SO_4^- ions which are soon taken up by plants. Some K^+ ions are also fixed in crystal lattice of clay and some are adsorbed. SO_4^- adsorption is more in calcareous soil so its use in calcareous and alkaline soils is very useful. K_2SO_4 has better effect in moist soil. It is a neutral fertilizer. Reaction in soil occurs as below-

$$\underset{\text{Fertilizer}}{K_2SO_4} \longrightarrow \underset{\text{Soil Solution and easily and soon taken up by plants)}}{2K^+ + SO_4^{2-}}$$

Effect on plants or crops: It is very soon available to plants. It increases the quality and quantity of crop by supplying S beside K. It is a very good fertilizer for tobacco, because tobacco leaves are of better quality containing excess of K without Cl. It is also used in tea and coffee garden to improve that quality.

Time and Methods application: K_2SO_4 should be added in soil before sowing or at the time of sowing. It can easily be mixed with ammonium chloride, super phosphate, rock phosphate and dicalcium phosphate and can be stored for some period. But after mixing with other fertilizers, it should be added in soil at once.

Potassium Magnesium Sulphate (K_2SO_4. 2Mg SO_4): After grinding Langbeinite ore chloride salts are separate from it by regulated running water. The Langbeinite remains as residue which is filtered and dried. This is used as potassium magnesium sulphate fertilizer.

Properties

Colour	Pink gray
State	Crystalline
Molecular weight	514.01
Melting point	926.7 ° C
% K	18.25
% K_2O	22.0
%MgO	10.85
Solubility	Soluble in water

Effect on Soil and Plants or Crops: This is an effective fertilizer in laterite soil. It can also successfully be used in sand loam and calcareous soils deficient in Mg. This fertilizer is generally used in all types of crops. It is efficiently used in Sugarcane and Pea crops.

Potassium Nitrate (KNO_3): It is also called Nitre, Salt-petre and Shora. It is a complete fertilizer but due to high cost per unite its use is very less. The reason of high cost per unite its main use in other industries such as medicine, chemical, explosive and fireworks. In industries it is used on the place of sodium nitrate. Hence, its use as fertilizer is negligible.

Potassium Carbonate (K_2CO_3): Wood ash contains potassium carbonate. Wood ash is directly used as manure in soil but K_2CO_3 can be separate by extracting wood ash. Extracting of K_2CO_3 from ash is difficult; hence, ash is directly used as manure in soil. Wood ash contains 0.5 % to 35 % K_2O. Very excess of potassium is present in tobacco ash.

Manufacturing

i. Synthetic method: By flowing air (CO_2) in potassium Hydroxide (KOH).

$$2KOH + CO_2 \longrightarrow K_2CO_3 + H_2O$$

By centrifuging the concentration solution, crystalline, K_2CO_3 is obtained.

ii. By flowing CO_2 after flowing current in potassium chloride (KCl) solution.

$$KCl \longrightarrow K^+ + Cl \qquad 2H^+ + 2e^- \longrightarrow H_2\uparrow(\text{Cathode})$$

$$HOH \longrightarrow H^+\ OH^- \qquad 2Cl^- - 2e^- \longrightarrow Cl_2\uparrow(\text{Anode})$$

$$2K + 2OH \longrightarrow 2KOH \qquad 2KOH + CO_2 \longrightarrow K_2CO_3 + H_2O$$

This fertilizer is not in common use. Its solubility is also less. In its place wood ash or ash can is used.

The use of ash is very useful in fruit plants, potato, vegetables etc.

Form of potassium absorbed by plants: Plant absorb potassium in the form of Potassium ion, K^+.

Fates of muriate of potash in soils

When added to soils, muriate of potash, KCl, dissociates to yield potassium and chloride ions in soil solution,

$$KCl \rightleftharpoons K^+ + Cl^-$$

A portion of K^+ is absorbed by plant roots. A portion is assimilated by soil microbes (immobilization). A portion may be leached with percolating water and another portion is fixed by the silicates clays.

Mixed Fertilizers (Fertilizer Mixtures): Mixture of two or more fertilizers, which supplies two or more nutrients, is called mixed fertilizers

Methods of Preparing Fertilizer Mixture: Fertilizers mixtures may be prepared by cultivars or in factories with the help of special equipment. The former are known as Farm mixed fertilizers while the letter is termed as factory made mixture.

1.Farm-Mixed Fertilizers or Mechanical Method: In this method, farmer makes mixed fertilizer himself at his farm or at his home. According to the need of crop and soil fertilizers are selected. The fertilizers are mixed thoroughly after spreading on a cemented floor. In this method, any instrumental or special technique is not required. Such types of mixed fertilizers are prepared at the time of sowing. In the form of conditioner dry soil or sand is mixed. In such type of fertilizers, caking does not occur and less cost is laid.

Advantages of Farm Mixed Fertilizers:

1. Caking does not occur hence, it can easily be applied in soil.
2. Farm mixed fertilizers are relatively cheap because no any special instrument of technique is required in its preparation.
3. Farmer can observe the effect and can change in the components of mixed fertilizers in future.

2.Factory-Made Mixture: **(Chemical Methods**): Four types of plants are used to prepare factory mixed fertilizers.

- Dry mix plant: Dry mix fertilizers are mixed by instruments. Dry mix plants are highly prevalent in India.
- Wet mix plant: Dry phosphatic and potassic fertilizers are mixed with liquid ammonia. This is adopted in USA.
- In this method, super phosphate is prepared by preparing sulphuric acid which is later on used in preparation of mixed fertilizers.
- In this method, super phosphate is prepared by general method which is mixed in fertilizer mixture.

Advantages of granulated Mixed Fertilizers:

1. Excess amount of nutrients are present in mixture.
2. There is no possibility of adulteration.
3. Homogeneous mixture is formed with all nutrients.
4. Fertilizer is not lost as dust in granular mixture.
5. The physical condition of the mixture is better.
6. More N, P and K are available to plants. Besides this, some micronutrients are also present.

Precautions in Making Fertilizer Mixture or (Incompatibilities in Fertilizers mixture:

a. Fertilizers Containing ammonia (*e.g.* AS AN) should not be mixed with basically reactive fertilizers(lime basic slag rock phosphate calcium cyanamide) as losses of nitrogen may result through escape of gaseous ammonia.

b. All water soluble phosphatic fertilizers like super phosphate and ammonium nitrate should not be mixed with those fertilizers that contain free lime converts a portion of the soluble phosphate into an insoluble form.

c. Easily soluble and hygroscopic fertilizers like calcium ammonium nitrate, urea and potash salts tend to cake or form lumps after mixing. Such fertilizers should be mixed shortly before use.

d. Slightly acidic super phosphate may liberate acid from certain fertilizers. (e.g. nitrate and chloride), and this may damage the gunny bags.

Quality Control of Mixed Fertilizers: From very beginning unfairness in the use of mixed fertilizers is going out. It is generally found that the materials or nutrients labeled on the sacks or bags are not present in that. Most of the mixed fertilizers making companies write the excess amounts of nutrients on the bags. For the controlling the quality of mixed fertilizers Government of India has fixed the purity limit of mixed fertilizers sample under the section 3 of essential commodity Act 1955 (Fertilizer Control Law 1957). Under this law, clear instructions have been given for registration of mixed fertilizer business, Chemical analysis of mixed fertilizer and printing of Grade on bags. On getting deficiency in the quality of mixed fertilizers there is a provision of punishment in the law.

Commercial Complex Fertilizers: When fertilizers supply only one plant nutrient, these are sometimes called "Straight fertilizer". For the last few years, new types of commercial complex fertilizers containing at least two or more of the primary essential nutrients (N, P and K_2O) have been manufactured on a very large scale in the USA and the European countries. Such fertilizers are called complex fertilizer.

The chief characteristics of such fertilizers are –

1. They usually have a high content of plant nutrient, more than kg per kg of fertilizers. As such they are also called "High analysis fertilizers".
2. They usually have a uniform grain size and good physical condition.
3. They supply nitrogen and phosphorus in an available form to the soil in one operation. Nitrogen is present as nitrate and ammoniacal form, and phosphorus is present in water soluble form up to 50 to 90 % of the total P_2O_5.
4. Since complex fertilizers are high analysis fertilizers, transport and distribution costs to the cultivators are less when worked out on the basis of per kg of plant nutrients.
5. They are non- caking and non- hygroscopic, thus, safer for storage.

Commercial Complex fertilizers being manufactured in India are-

1. Nitro phosphates also known as Nitric-Phosphate and
2. Ammonium phosphate sulphate, Diammonium phosphate and urea ammonium phosphate.

1. Nitro phosphates: There are many processes by which nitro phosphates are manufactured. Out of which, the following processes are commercially important.

A. Process involving treatment of rock phosphate with HNO_3
   - Nitro- carbonic process: PEC or Carbonic nitro phosphate.
   - Nitro-Separation process: ODDA nitro phosphate.

B. Process involving use of mixture of Acids
   - Process involving use of nitric acid and phosphoric acid: Phosphoro-nitric, Nitro-phosphate
   - Process involving use of nitric acid and sulphuric acids: Sulpho-nitric, Nitro-phosphate

PROCESS INVOLVING TREATMENT OF ROCK PHOSPHATE WITH HNO_3

A. Nitro-carbonic process: PEC or Carbonic nitro phosphate: Forty mesh rock phosphates are treated with nitric acid and this mass is later treated with NH_3 and CO_2 in the presences of a stabilizer with $MgSO_4$. A stabilizer is added to prevent reversion of citric soluble phosphate to insoluble phosphate.

Acidulation: First stage

$$\underset{\text{Rock phosphate}}{Ca_{10}F_2\,(PO_4)_6} + \underset{\text{Nitric acids}}{20\,HNO_3} \rightleftharpoons \underset{\text{Phosphoric acids}}{6H_2PO_4} + \underset{\text{Calcium Nitrate}}{10\,Ca\,(NO_3)}$$

D.A.P. addition: Second stage (addition of water soluble phosphate)

$$+\,7\,(NH_4)\,HPO_4 \rightleftharpoons \underset{\text{Ammonium nitrate}}{14\,NH_4NO_3} + \underset{\text{Calcium nitrate}}{3\,Ca\,(NO_3)} + \underset{\text{Mono calcium phosphate}}{6Ca\,(H_2PO_4)} + \underset{\text{Calcium fluoride}}{CaF_2} + \underset{\text{Phosphoric acid}}{H_3PO_4}$$

Ammoniation Third stage:

$$+\,10NH_4 \rightleftharpoons \underset{\text{Ammonium nitrate}}{20\,NH_4NO_3} + \underset{\text{Monocalcium phosphate}}{4NH_4\,H_2PO_4} + \underset{\text{DCP}}{9\,CaHPO_4} + \underset{\text{Calcium fluoride}}{CaF_2}$$

B. Nitro-Separation process: ODDA nitro-phosphate:

This process involves acidulation of rock phosphate with HNO_3, partial separation of $Ca(NO_3)$ by cooling and crystallization, ammoniating the phosphate solution to convert the NO_3 to NH_4NO_3 and PO4 to $CaHPO_4$ and drying the ammoniated slurry to produce an N-P fertilizer typically of 20-

22 grades. $Ca(NO_3)$ is dehydrated and either marked as such or ammoniated and carbonated to yield calcium ammonium nitrate to contain 25.0 % Nitrogen.

PROCESS INVOLVING USE OF MIXTURE OF ACIDS:

Phosphoro-nitric Nitro-phosphate: In this process, rock phosphate is treated with nitric acid and sufficient quantity of phosphoric acid to adjust the molecular ratio of CaO: P_2O_5 to 2 or less as compared with a ratio of about 3.5 to 4 in rock phosphate. The following reactions are take place;

$$\underset{\text{Rock phosphate}}{Ca_{10}F_2(PO_4)_6} + \underset{\text{Nitric acids}}{16\,HNO_3} \rightleftharpoons \underset{\text{Phosphoric acid}}{4H_2PO_4} + \underset{\text{Phosphoric nitrate}}{6\,NH_3PO_4}$$

$$+ \underset{\text{Monocalcium phosphate}}{2Ca(H_2PO_4)_2} + \underset{\text{Calcium Nitrate}}{8Ca(NO_3)} + \underset{\text{Hydrogen Fluoride or Hydrofluoric acid}}{2HF}$$

In this process, calcium nitrate is formed which make the slurry or treated mixture, unsuitable for direct use as a fertilizer. The total plant nutrient content of nitro-phosphate manufactured by this process ranges from 39 to 47%.

Sulphonitric Nitro phosphate: In this process rock phosphate is treated with a mixture nitric acid and sulphuric acid. The following reactions are take place;

$$\underset{\text{Rock phosphate}}{Ca_{10}F_2(PO_4)_6} + \underset{\text{Nitric acids}}{12\,HNO_3} + \underset{\text{Sulphuric Acid}}{4H_2SO_4} \rightleftharpoons \underset{\text{Phosphoric acid}}{6H_3PO_4} +$$

$$\underset{\text{Calcium sulphate}}{4Ca\,SO_4} + \underset{\text{Calcium Nitrate}}{6\,Ca(NO_3)_2} + \underset{\text{Hydrogen Fluoride or Hydrofluoric acid}}{2HF}$$

The slurry is eliminated and treating with liquid ammonia. The following reactions take places:

$$\underset{\text{Phosphoric acid}}{6H_3PO_4} + \underset{\text{Calcium sulphate}}{4Ca\,SO_4} + \underset{\text{Calcium Nitrate}}{6\,Ca(NO_3)_2} + \underset{\text{Hydrogen Fluoride, Hydrofluoric}}{2HF} + \underset{\text{Liquid or ammonia}}{NH_3}$$

$$\rightleftharpoons \underset{\text{Dicalcium Phosphate}}{5CaHPO_4} + \underset{\text{Monoammonium phosphate}}{NH_4H_2PO_4} + \underset{\text{Calcium sulphate}}{4\,CaSO_4}$$

The total plant nutrient content of nitro phosphate manufactured by this process ranges from 26 to 33 % only.

Characteristics and use of Nitro phosphate: Nitro phosphate is a granulated fertilizers containing stabilizer which prevents reversion of citrate soluble phosphate to insoluble phosphate. Being granulated, it maintains excellent physical condition during storage and handling. All types of nitro phosphates contain easily soluble and readily available nitrate nitrogen (NO_3-N) as well as readily available ammoniacal nitrogen (NH_4-N). In Sulphonitric nitrophosphate, phosphorus part is present in water soluble

gypsum and in acidic soil as lime. Due to excess use of high grade fertilizers and prevalence of sulphur less fertilizers now-a-days deficiency of S is common. So the use of S has become essential like N,P and K fertilizers. S is also used as soil amendment in alkaline soil.

Secondary Nutrient

1. *Calcium:* Source of soil Ca: The Ca concentration of the earth's crust is about 3.64 %; however, the Ca^{2+} content in soils varies widely. Sandy soils of humid regions very low amounts of Ca^{2+}. The plasioclase mineral anorthite ($CaAl_2Si_2O_3$) is the most important primary source of Ca. The Ca content of arid region soils is generally high as a result of low rainfall and little leaching. Calcite, ($CaCO_3$) is often the dominant source of Ca in soils of semi arid and arid regions. The fate of solution Ca^{2+} is less complex than that of K^+. As a general rule, coarse textured humid region soils formed from rocks low in Ca minerals are low in Ca. Soils that are fine textured and formed from rocks high in the calcium mineral. Calcium occurs in nature in carbonate, sulphate and hydroxide forms complex calcium silicates and organic matter. It occurs as exchangeable cation in the vast majority of approximately neutral or slightly saline soils. Such soils are located in the Deccan tract of India mainly as deep black and medium black soils and in northern India as alluvial soil. The acid soils of India are low in calcium. Thus, the low pH of acid soils is indicative of calcium deficiency. The neutral and slightly alkaline soils of India are rich in Ca; there is no special need to supply calcium to these soils. Percentage of Ca in commonly used fertilizers is given in bellows.

Fertilizers/ Soil amendment	% Ca	%Mg
Nitrogenous fertilizers		
Calcium nitrate	19.5	1.5
Calcium Ammonium Nitrate	8.1	4.5
Calcium Cyanamide	39.1	-
Phosphatic fertilizers		
Super phosphate (Single)	19.5	0.3
Super phosphate (Triple)	14.3	-
Bone meal	23.0	-
DCP	22.9	-
Basic slag	33.9	9.0
Potassic fertilizers		
Potassium sulphate	-	0.6
Potassium magnesium sulphate	-	10.8
Soil amendments		
Lime stone	32.3	4.0-12.0
Gypsum	29.2	-

About one-third of Indian soils are acidic. By virtue of having low base saturation, particularly in high rainfall areas, many acids soil may be deficient in Ca (Tandan, 1991). However, soil containing less than 25% of their cation exchange capacity (CEC) or less than 1.5 me exchangeable ca/

100 gm soil have usually been considered as Ca-deficient. *Legumes are generally more responsive to Ca than cereals.*

2. *Magnesium:*

Form utilized by plants: Magnesium (Mg) is absorbed by plants as Mg^{2+} from the soil solution.

Sources of Soil Mg: Mg constitutes 1.93% of the earth's crust; however the Mg^{2+} content of soils is variable. Vermiculture has a high Mg content.

Hypomagnesemia or Grass tetany: Cattle consuming low Mg forage may suffer from Hypomagnesemia or Grass tetany which is an abnormally low level of blood Mg. This essential plant nutrient has not been observed to be deficient on a wide scale in any part of India. However, usually soils containing less than 1 me exchangeable Mg/100 gm soil or less than 4-5 % of CEC occupied by Mg are considered deficient (Biswas et.al.1985). Magnesium is indirectly applied to the soil through commercial fertilizers.

3. *Sulphur:*

Sources in soil: Sulphur is the 13th most abundant element in the earth's crust, averaging between 0.06 and 0.10%. The main S- bearing mineral in rocks and soils are Gypsum, ($CaSO_4$. $2H_2O$) sphalerite (Zn S) epsomite ($MgSO_4$. $7H_2O$) Pyrites (Fe S) etc. S can be much more abundant in Biotite, chlorite and layer types clay mineral.

Forms of S in Soils: S is present in the soil in both organic and inorganic forms, although nearly 90% of the total S in most non calcareous surface soils exits in organic forms. The inorganic forms are solution SO_4^{2-}, absorbed SO_4^{2-}, insoluble SO_4^{2-} and reduced inorganic S compound. Sulphur is considered as the forth major nutrient for plant growth. Crops generally absorb S in amount comparable to P which is another major nutrients close to 25% of the total cultivated area is estimated to be suffering from S deficiency.

Fertilizers	Formula	% S
Ammonium sulphate	$(NH_4)_2SO_4$	24.0
Ammonium sulphate Nitrate	$NH_4H_2SO_4$. NH_4NO_3	15.0
Super phosphate single	$Ca(H_2PO_4)_2.2CaSO_4.H_2O$	12.0
Potassium Sulphate	K_2SO_4	18.0
Ferrous sulphate	$FeSO_4$	19.0
Copper sulphate	$CuSO_4.5H_2O$	13.8
Elemental Sulphur	S	80-100
Manganese Sulphate	$MnSO_4.5H_2O$	17.0
Magnesium sulphate	$MgSO_4$	23.0
Zinc sulphate	$ZnSO_4.6H_2O$	11-18
Gypsum	$CaSO_4.2H_2O$	13-18

It is essential part of living cell. It plays a great role in protein formation and thus, improves the quality of crop yield. Its importance is more in leguminous and oilseed crops. It is supplied in soil indirectly because a number of fertilizers having S are added in soil for supplying other nutrients. A few S containing fertilizers are as follows.

Sulphur and Crop Quality: Sulphur plays an important role in improving the quality and marketability of the produce. Sulphur can improve crop quality in the following ways.

1. Increasing the oil content of seeds.
2. Synthesis of S-containing amino acids and protein percentage.
3. Improving nutritional quality of forages.
4. Improving starch content of tubers.
5. Improving baking quality of wheat.
6. Raising sugar recovery in sugarcane.
7. Enhancing marketability of copra.

Economic of Sulphur Application: At the present fertilizer prices is the cheapest of the major nutrients, being Rs. 3.65 per kg of S compared to Rs. 6.00 per kg of N or Rs. 11.80 per kg of P_2O_5 (Tandon,1995).

Source of Sulphur: Over 60 fertilizers containing S are known to agriculture but main fertilizers and soil amendments as source of S in India are given as follows: Sulphur content of common fertilizers and soil amendments consumed in India

Fertilizer/ soil amendment	S%
Ammonium sulphate	24
Ammonium sulphate nitrate (ASN)	15
Single super phosphate	12
Ammonium Phosphate Sulphate	15
Potassium Sulphate	18
Gypsum	13-21
Pyrite	22
Zinc sulphate	11-18

Besides addition of S through fertilizers containing Sulphur, it is also advisable to apply 20-50 kg S / ha as elemental S mainly to oilseed crops. This dose should be applied 3-4 week before planting in most and aerated soil.

Mineralization of S-compound in soil

The sulphur in soil is being cycled continuously between inorganic and organic soil form. The nature of compounds formed and their transformations are influenced by the biologically mediated processes, which, in turn, are affected by environmental conditions. As has been stated earlier in many cases, organic matter is the major source of soil sulphur. Its oxidation to SO_4^{2-} is brought about by soil microorganism and the process is called mineralization. It is this mineralized form of S which can be taken up by plants and microorganisms. Rate of mineralization is affected by factors such as moisture, aeration, and temperature and soil pH. The conversion of organic–S to SO_4^{2-} is an oxidation process and can not take place in the absence of oxygen. Proper aeration is, therefore, essential for this process to proceed. Also, this process can not take place under submerged soil condition except in the aerated pocket of the soil. The mineralization process can be expressed by following equation-

$$\text{Organic S} \xrightarrow[O_2]{\text{Aeration}} SO_4^{2-} + H^+$$

This formation of end product SO_4^{2-} is associated with the formation of H^+. Hence; this process lowers the soil pH. This biological transformation in some cases requires very low pH in soil before the process can be initiated. Under the limited S supply and in the presence of excess of carbonaceous materials provide energy to the growing population of microorganism, which consume the mineralized S. Hence, crop plants may suffer from S deficiency. This process is called "immobilization". Immobilization is a temporary phase as on the death of the microbial population the microbial S is mineralized to SO_4^{2-} which can be utilized by growing plants. Sulphur in the soil is associated with organic carbon in a fixed C: S ratio of about 140:1. Thus, a favourable C: S ratio is more important than a large store of organic matter for S response under tropical conditions. This is because the ratio influences microbial decomposition of organic matter. A major transformation involving S is that of conversion/ oxidation of organic–S into sulphate. This process is similar to N-transformation optimum Since, this transformation is mediated biologically, and extremes of temperature are 35-40°C. Soil moisture is another important factor. Which moisture level of about 60% of field capacity is good for oxidation of organic –S to sulphate excessive moisture decrease the rate of oxidation due to low accessibility of oxygen or aeration.

Importance of Secondary Nutrients in Agriculture: The importance of Ca and Mg attain importance in areas where soils are either strongly alkaline or acidic. These soils are often deficient in Ca and Mg. Alkali soils occupy about 2 million hectares of land in India. These soils have high exchangeable Na; and Ca is a suitable ion to replace this Na from the exchange complex. Ca is supplied through gypsum. The replaced sodium forms sodium sulphate which is leached down. Gypsum also reacts with sodium carbonate to form CaCo3 and NA_2SO_4 which is leached down the profile. A through leaching of the soil with irrigation water should be done after application of gypsum to free it from sodium sulphate. Acid soil, calcium saturation of exchange

Crop response to liming: Judicious use of liming increases the yields and improves the efficiency of fertilizers in acids soils.

Liming Materials: The most common agricultural liming materials are ground limestone, burnt lime and hydrated lime. The chemical relationship of these there is shown in the equation:

$CaCO_3$ (Lime stone) CaO (Burnt Lime) Ca $(OH)_2$ (Hydrated Lime)

Of the three forms of liming materials ground lime stone is most extensively applied to the soil in various countries.

When lime is applied to an acidic soil, a part of calcium carbonate undergoes solution and combines with carbonic acid to form calcium bicarbonate. Calcium bicarbonate in solution form reacts with the soil colloids freely evolving carbon dioxide.

$$CaCO_3 + H_2CO_3 \longrightarrow Ca\ (HCO_3)_2 \qquad .....1$$

$$\text{Micelle} + Ca\ (HCO_3)_2 \xrightarrow[H]{H} \text{Ca Micelle} \qquad .....2$$

While some portion of the lime stone goes into solution a part of the applied lime stone comes in close contact with the soil colloids in solids conditions. Under these circumstances the absorption of calcium with soil colloids also occurs as indicated below.

$$\text{Micelle} + CaCO_3 \xrightarrow[H]{H} \text{Ca Micelle} + H_2O + CO_2 \qquad ...3$$

As the above reactions of lime stone proceed carbon dioxide is freely evolved and the absorption of the calcium and magnesium raises the percentage base saturation of colloidal complex correspondingly pushing up the pH of the soil solution.

The quality of ground lime stone and its value as soil amendment depends on both its chemical composition and its degree of fineness.

By product lime: Besides, the natural source of lime known as ground or crushed limestone liming materials also produced as by products in the paper sugar and ammonium sulphate industries. These are chemically precipitated calcium carbonate in fine form. The Sindri Fertilizer Factory is at present producing lime as a by product in the production of ammonium sulphate by the gypsum process.

Burnt lime: It is also known as quick lime oxide of lime, caustic lime or stone lime. When lime stone is heated to over 8000°C the calcium carbonate decomposes, yielding calcium oxide as shown in the following equations.

$$CaCO_3 \xrightarrow[8000°C]{Heat} CaO + CO_3$$

For every 100 kg of calcium carbonate burnt, 56 kg calcium oxide is formed. In the same way, for every 84.3 kg of magnesium carbonate burnt 40.3 kg of magnesium oxide is formed.

The purity of burnt lime sold for agricultural purpose is variable ranging from 85 to 98 %. When burnt lime is added to any acid soil, it changes to bicarbonate form as it goes into solution according to the following equation.

$$CaO + H_2O \longrightarrow Ca\ (OH)_2$$

$$Ca\ (OH)_2 + 2H_2CO_3 \longrightarrow Ca(HCO_3)_2 + 2H_2O$$

Calcium bicarbonate thus, formed reacts with the soil colloids and calcium ions are absorbed in then colloids complex.

$$\text{Micelle} + Ca\ (HCO_3)_2 \xrightarrow[H]{H} Ca\ \text{Micelle} + 2H_2O + 2\ CO_2$$

Slaked lime: It is also known as hydrated lime or hydroxide of lime. It is produced by adding water to burnt lime as:

$$\underset{\text{(Burnt lime)}}{CaO} + \underset{\text{(Water)}}{H_2O} \longrightarrow \underset{\text{(Slaked lime)}}{Ca\ (OH)_2}$$

For every 56 kg of calcium oxide, 18 litre of water are needed for proper slaking to yield 74 kg of hydrated or slaked lime. During this process a considerable amount of heat is generated. The final product marketed as agricultural lime is a whitish–grey or grayish-white powder which is caustic and alkaline.

When slaked lime is added to an acid soil, it combines with carbonic acid to form calcium bicarbonate in soil complex as:

$$\underset{\text{(Slaked lime)}}{Ca\ (OH)_2} + \underset{\text{(Carbonic acids)}}{2H_2CO_3} \longrightarrow \underset{\text{(Calcium bicarbonate)}}{Ca\ (HCO_3)_2} + \underset{\text{(Water)}}{2H_2O}$$

$$\text{Micelle} + Ca\ (HCO_3)_2 \xrightarrow[H]{H} \underset{\text{(Soil Colloids)}}{Ca\ \text{Micelle}} + 2H_2O + 2\ CO_2$$

Comparative Efficiency of Liming Materials: All forms of lime mentioned above are not of equal value for correcting soil acidity. The effectiveness of efficiency of any liming materials depends on the following factors:

1. The chemical composition of the material expressed in term of calcium carbonate equivalent known as neutralizing value (N.V.)
2. **The purity of the material:** the purer the material, the more effective it is for neutralizing acidity.
3. The degree of fineness of the materials.

1. Neutralizing Value: Pure calcium carbonate is taken as the standards against which other liming materials are measured. The neutralizing value of $CaCO_3$ is considered to be 100 %, $CaCO_3$ equivalent or N.V. of burnt lime or CaO can be worked out as follows.

$$CaCO_3 \text{ equivalent of } CaO = \frac{\text{Molecular weight of } CaCO_3}{\text{Molecular weight of } CaO} \times 100$$

$$CaCO_3 \text{ equivalent of } CaO = \frac{100}{56} \times 100 = 178.6$$

The same procedure can be applied to calculate the neutralizing value of other liming material. The neutralizing values of commonly used liming materials are given below.

Neutralizing values of the pure forms of commonly used liming materials

Materials	Chemical Composition	Neutralizing Value
Lime stone	$CaCO_3$	100 %
Burnt lime	CaO	179 %
Slaked lime	Ca (OH_2)	136 %
Dolomite	Ca Mg $(CO_3)2$	109 %

2. Purity of the material: The effectiveness of a liming material depends on its purity from the composition point of view e.g. if ground lime stone contains 95 % of $CaCO_3$, the NV of the material will be 95%.

3. Degree of fineness of the material: The degree of fineness is equally important because the speed with which a liming material reacts with an acid soil depends upon its fineness. The finer the material is ground the more rapidly it goes into solution and is more and is more effective. The fineness is measured in terms of the ability of a materials to pass through a sieve having 60 holes of equal size in one linear inch. Such a sieve is called a 60–mesh sieve and material passing through such a sieve is allotted 100 % efficiency rating. The scale of measuring the efficiency of liming materials on the basis of its fineness has been defined differently in various parts of the World.

Amount of lime to supply: The rates of liming would depend upon the following factors.

- **pH or intensity of the soil acidity**: For strong acid soils more liming material will be required compared to slightly acidic soil.
- **Cation Exchange Capacity of the soil**: A specific lime required of each soil depends upon its cation exchange capacity which is controlled by the amounts and types of clay and organic matter contained. The grater the cation exchange capacity, the greater the amount of lime required to raise the pH from value to another.
- **Texture of the soil**. Sandy soils require less lime compared to heavy soils, like silt loam and clay loam.
- **Lime requirement of crops in included in rotation**: Pulse and legumes benefit most by the application of lime since they do not thrive on

strongly acid soils and are also large consumers of lime. For the potato crop lesser quantity of lime should be applied.

- **Purity of the liming material**: The purer the material, the lesser the quantity required.
- **Chemical composition of the liming materials:** Since the neutralizing value of calcium carbonate is taken as 100, the amount of burnt lime required for application will be less.
- **Fineness or particle size:** The finer the liming materials, the more effective will it is.

Methods of applying lime: The following general principles should be observed in applying lime. Lime should be applied before ploughing or applied on ploughed land then disced or mixed into the soil.

When large quantities of lime are to be applied, it is advisable to apply small quantities two or three times rather than giving one heavy application. On strongly acid soil, where three to six tonnes of lime are required, half the quantity should be applied before ploughing and the remaining half applied and disced in after ploughing. When the time requirement is up to two tonnes per hectare, the entire amount should be applied and disced in at one time, 8 to 10 days before sowing the crop. when only one application of lime is to be given, it should never be given before ploughing with a furrow turning plough.

Amendments for reclaiming saline and alkaline soils: Some chemicals are added to the soil as an integral part of reclamation programme adopted to improve saline and alkaline soils. These are known as chemical amendments. The principal purpose of the amendments is to furnish soluble calcium to replace exchangeable sodium or to neutralize alkaline salts with acid.

Various chemical amendments suitable for different soil conditions are:

1. Gypsum ($CaSO_4 \cdot {}_2H_2O$)
2. Sulphur (S)
3. Iron sulphate ($FeSO_4$)
4. Lime stone ($CaCO_3$)
5. Pyrites (FeS_2)

1. Gypsum ($CaSO_4$): Gypsum a natural sulphate of calcium is found in large deposits in various parts of Rajasthan. It reacts with exchangeable sodium which gets converted into sodium sulphate. Sulphate is leached from the soil to reduce the soil pH. The following chemical reaction takes place.

$$CaSO_4 + 2\ Na\ X \longrightarrow Ca\ X + Na_2SO_4$$

The rate of application of gypsum is usually one to two tonnes per hectare. But it can be safely applied even up to 20 tonnes per hectare, depending on the soil pH and the concentration of exchangeable sodium present in the soil complex. Bhumbla (1974) reported that light soil requires less gypsum compared to medium or heavy soil.

2. Sulphur: Sulphur is a very effective chemical amendment to replace exchangeable sodium. Theoretically one atom of sulphur replaces four sodium ions by calcium. But under field conditions approximately three exchangeable sodium ions atoms of sulphur are replaced from the soil colloids. A chain of relations that take place when sulphur is added is.

$$2S + 3O_2 \longrightarrow 2SO_3 \text{ (microbial oxidation)}$$
$$SO_3 + H_2O \longrightarrow H_2SO_4$$
$$H_2SO_4 + CaCO_3 \longrightarrow CaSO_4 + CO_2 + H_2O$$
$$2Na\,X + CaSO_4 \longrightarrow CaX_2 + Na_2SO_4$$

The quantity of sulphur required per hectare to change the soil pH to about 6.5 would depend on the soil and climate conditions. As the soil pH increase, more of sulphur will be required to reduce the soil pH to about 6.5.

3. Iron sulphate: Iron sulphate is some times used as a chemical amendment for improving alkali soils. iron sulphate forms sulphuric acid which is converted into calcium sulphate calcium sulphate thus, formed replaces exchangeable sodium as indicated by the following equations:

$$FeSO_4 + H_2O \longrightarrow H_2SO_4 + FeO$$
$$H_2SO_4 + CaCO_3 \longrightarrow CaSO_4 + H_2O + CO_2$$

4.Lime stone ($CaCO_3$): Ground lime stone is applied to the soils having pH of 7 to 7.5. Since calcium carbonate becomes insoluble as the pH increases. It is not effective on soils having pH more than 7.5. The following reactions take place with the soil complexes.

$$2Na\,X + CaCO_3 \longrightarrow Ca\,X + Na_2CO_3$$

When lime stone is applied to the soil, it gets dissolved in they soil solution. The calcium part of the lime stone reacts with the soil complex and replaces sodium. Sodium combines with carbonate cations to form sodium carbonate which is leached out of the soil flooding.

5. Pyrites: Iron pyrites is a mineral containing iron and sulphur. Because of the acidifying characters, useful ness of pyrites in the reclamation of alkali soils has been established in practice. Pyrites after application on the moist surface soil undergoes a series of chemical and partly microbial oxidation reactions with the formation of sulphuric acid. The overall chemical reactions are as follows:

Pyrites + Oxygen + Water $\longrightarrow$ Sulphuric acid
(Sulphur)

Alkali soils + Sulphuric acid+ Soil – Calcium Carbon di oxide + water
(Calcium carbonate)

Soil- sodium + calcium sulphate (Sodic –Clay) → Soil Calcium + Sodium sulphate (Calcium –Clay) (Leached) Fertile soil

So, the sulphur in pyrites gets oxidized to sulphuric acid which reacts with calcium carbonate present in the alkali soil and forms calcium sulphate. This calcium in turn replaces excessive sodium from the soil exchanges complexes and reclaims the alkali soil.

All chemical amendments listed below are not suitable of all soil conditions the amendments suitable for different soil conditions are:

Chemical amendments		Suitable for
Gypsum	::	Saline and alkaline soils having pH up to 9.0
Sulphur	::	Alkaline and Saline-alkali soils having pH up to 9.0
Iron sulphate	::	Alkaline and Saline-alkali soils having pH ranging up 9.0
Lime stone	::	Saline soils having pH less than 8.0
Pyrites	::	Sodic soil with pH less than 8.0

Amongst these five amendments sulphur is most effective in improving saline and alkaline soils. One tonnes of sulphur is equivalent to —

- 5.38 tonnes of gypsum
- 6.64 tonnes of iron sulphate and
- 3.13 tonnes of lime stone

Soil conditioner: Soil conditioner is materials which are added to improve or maintain physical condition of the soil. Crop residue, manure and other organic substances have long been recognized as excellent materials for improving the soil structure or physical condition of soil, thereby increasing the infiltration of water or reducing the run off. In recent years mainly after 1950 synthetic organic materials have been produced forming and stabilizing the soil aggregates. These materials are poly electrolytes including polyvinyl elites, polyacrylates, cellulose gums, lignin derivatives and silicates.

Since, these materials do not develop a structure but merely preserve the soil against deterioration; they are plied to the soil when the soil exhibits a good structure. Soil properly treated with effective soil conditioners may maintain a good structure for ma number of years. Soil conditioners are used wit advantages in stabilizing the soils on slopes subject to rapid erosion.

At present, soil conditioners are very costly and as such are not used on field scale. However, for high value green house and market garden crops, these are used in the USA and European countries.

Placement and application of fertilizers: The main aim of fertilizer application is to obtain maximum yield unit^{-1} area minimum cost and labour. For obtaining maximum yield it is very essential to decide the type and amount of fertilizers, time and method of application and the type of crop to be taken. Only those nutrients as fertilizers should be added with are deficient in soil. Fertilizers should be added in soil in such a way that sufficient amount of nutrients could be efficient use of fertilizers is adopted to achieve adopted objectives.

Principles of efficient use of fertilizers: The principles of efficient use of fertilizers are 1) selection of fertilizers 2) proper time of application of fertilizers 3) use of suitable amount of fertilizers 4) correct method of application of fertilizers.

1. Selection of fertilizers: Correct selection of fertilizers depends on the following factors.

(i) **Physical and chemical properties of soil:** Due to excess leaching in light soil nitrate fertilizers should not be added. Fertilizers producing acidity as ammonium sulphate should not be added in acidic soil. Water soluble phosphatic fertilizers as super phosphate should not be added in acid soil because they are readily fixed in soil. Alkalinity producing fertilizers like sodium nitrate should not be added in alkali soil.

(ii) **Climate**: In high rainfall area ammonical fertilizers should be added in soil because the leaching loss is less, whereas nitrate fertilizers can easily be applied in soil where rainfall is low.

(iii) **Crop requirement**: Ammonical nitrogen is better than nitrate nitrogen in paddy crop where as some crop like tobacco, linseed etc are very sensitive to chloride; hence fertilizers containing chloride should not be added in soil. When crops require excess sulphur i.e. ground nut oilseed crops, pulse crops etc sulphur containing fertilizers like gypsum, ammonium sulphate should be added in soil.

(iv) **Unit cost of nutrients**-The fertilizers having minimum nutrient unit cost value should be applied in soil.

2. Proper time of application of fertilizers: Following factor affect the correct time of application of fertilizers-

(i) **Soil type**: Nitrogenous fertilizers should be added in two or three splits doses in soils having less water holding capacity.

(ii) **Duration of crop:** For long duration of crops fertilizer should be added in soil in several slit doses.

(iii) **Weather effect:** In rainy season nitrogen fertilizers should be applied in several splits doses where as in dry season spray or foliar application of fertilizers is better.

(iv) **Kind of fertilizers:** Nitrogenous fertilizers should be applied in several split doses whereas phosphorus and potassium should be applied in soil before sowing or at the time of sowing. In light soil potassium may be applied in several split doses. Micronutrient

fertilizers should be applied at the time of sowing they should be sprayed in standing crop.

3. Amount of fertilizers: Quantity of fertilizers to be applied is highly affected by the following factors.

(i) **Crop requirement:** Different crop requires different amount of fertilizers besides, different amount of fertilizers is needed for different for different variety of a crop. High yielding varieties need relatively excess amount of fertilizers than low yielding varieties.

(ii) **Objective of growing crop:** Crop grown for fodder requires fewer amounts of fertilizers than grown for grain.

(iii) **Amount of nutrients present in soil:** By soil testing methods available amount of nutrients in soil are determined.

(iv) **Crop rotation:** If best crop rotation is adopted fewer amounts of fertilizers is required. In crop rotation leguminous crops and green manuring crop should be included.

(v) **Soil erosion:** Soil fertility is less of eroded soils. In such type of soil excess amount of fertilizers is required. Generally in such soil nitrogenous fertilizers are required more because loss of nitrogen is more.

(vi) **Soil pH:** Soil pH highly affects soil fertility. In highly acidic and alkaline soil availability of phosphorus is less; hence excess amount of P fertilizers is required.

(vii) **Soil moisture:** In dry soil fewer amounts of fertilizers should be added otherwise on applying excess amount of fertilizers adverse effect occurs on germination plant growth and yield of crops.

4. Correct method of fertilizers application: Methods of applying fertilizers depends on the following factors-

(i) Types, availability amount and unit^{-1} cost of fertilizers.

(ii) Amount of soil moisture

(iii) Soil types

(iv) Climate

(v) Methods of sowing

(vi) Type and age of crop

(vii) Availability of labour and capital.

Nitrogenous fertilizer should be applied by broadcasting method. In standing crop they can be applied as top dressing. Nitrogenous fertilizers can also be applied as spray or foliar application.

Phosphatic fertilizers should be applied in furrow. The best method of applying phosphatic fertilizers is **localized placement** near seed because phosphorus is less mobile.

Potassic fertilizers should also be used below near the root zone.

Different methods of fertilizers application: Following methods of applying fertilizers are prevalent.

(*i*) Application of fertilizers in solid form

(*ii*) Application of fertilizers in liquid form

1. Application of fertilizers in solid form:

I. Broadcasting Method

(*i*) Broadcasting before or at the time of sowing

(*ii*) Top dressing

A. Broadcasting in standing crop

B. Spot dressing

C. Side dressing

II. Localized placement

(*i*) Dibbling placement

(*ii*) Contact or drilling placement

(*iii*) Band placement

A. Row placement

B. Hill placement

C. Pellet placement

III. placement

(*i*) Deep placement

(*ii*) Sub-soil placement

(*iii*) Plough sole placement

I. Broadcasting Method: Spreading of dry and solid fertilizers on soil surface and mixing in soil by ploughing or sowing is generally called broadcasting. There is two types-

(*i*) **Broadcasting before or at the time of sowing**: In this method fertilizers are broadcasting in the field one day before or at the time of sowing and sowing is done. This method is very common in India. Half amount of nitrogen and total amount of phosphorus and potassium are applied by this method. Secondary and micro-nutrient fertilizers are also applied by this method. In hilly areas, broadcasting is done by aero plane.

(*ii*) **Top dressing:** Broadcasting of fertilizers in standing crop is called top dressing. This can be done in two ways-

A. Broadcasting in standing crop: Nitrate and sometime ammonical nitrogen is top dressed in standing crop to supply nutrient at once.

B. Spot dressing: Crops in which there is some distance between two plants, fertilizers are kept around the plant or in side near the plant. It is called spot dressing.

C. Side dressing: broad casting of fertilizer, in standing crop sown in lines in side of plants is called side dressing.

Advantages of broadcasting methods

(i) Less time is required in applying the fertilizers.

(ii) It is relatively less costly.

(iii) It is very conveniently in our country because most of the farmers are still illiterate.

Disadvantages of broadcasting methods

(i) Plants do not get the total advantage of fertilizer because some amount of fertilizer is beyond the reach of plant roots.

(ii) Weeds grown in the field take the benefits of fertilizers ultimately less amount of fertilizers is available to plants.

(iii) Some amount of fertilizers is fixed in soil.

II. Localized placement: Placing of fertilizers near the seed or plant at a special place is called local placement. Different methods is as below.

(i) **Dibbling placement**: Fertilizers are placed by dibbler instrument. By dibbler holes are made in field and seeds and fertilizers are added in the holes together. This method is costly and unsuitable because less germination occurs.

(ii) **Contact or drilling placement:** Sowing of seed and fertilizers together in furrow is called contact or drilling placement. It is common for the crops sown in lines. Application of seed and fertilizer separately is batter which is very useful for dry farming.

(iii) **Band placement:** Fertilizers placed in band in the side of plants. This method is used in those crops in which distances in between lines and sides more. Phosphorus and potassic fertilizers are used by this method. Maximum amount of fertilizers is utilized by plants in this method. It is different types like-

A. Row placement: This method is used for those crops which are sown in lines with less distance in between the lines. Fertilizers are applied in sugarcane, maize, cotton, potato, vegetable and cereal crops.

B. Hill placement: This method is used for those crops which are sown in line with more distance in between the lines (1meter or more). Fertilizers are applied by this method in fruit plants like guava, apple, grave, pear, orange, lemon, banana etc.

C. Pellet placement: In this method, pellets of mixture of fertilizers and moist soil are made which are placed near the plants 3-6 cm below the surface. Some times fertilizers is rolled in paper and used as pellet. This is the best methods for increasing the utility capacity of urea.

III. Placement: Keeping of fertilizers at a definite and suitable place in the field is called local placement. Plant roots are active in a definite area. So it is essential to place the fertilizers in that area to increase the efficiency of fertilizer. Placement are done in different ways-

(i) Deep placement: To control the loss of fertilizers it is kept in soil at more depth. As fertilizer is in anaerobic horizons its loss does not occur as ammonia gas and nitrogen is being regularly absorbed by plants. It is adopted in Japan and Philippines in paddy crops.

(ii) Sub-soil placement: This method is used in acidic soil of moist area. Phosphatic and potassic fertilizers are placed in sub soil by instruments.

(iii) Plough sole placement: Making furrow by plough fertilizers is placed in the bottom of furrow where most of the plant roots develop. Thus, they absorb the nutrients quickly. This method is also suitable for dry area.

Advantages Placement methods

(i) Plant nutrients become easily available because fertilizers are placed near plant root zone.

(ii) Fixation of phosphorus and potassic fertilizers is less.

(iii) Plants grow rapidly and weeds do not grow.

(iv) Germination is not affected by fertilizer placement.

(v) The residual effect of fertilizer remains far longer period.

Disadvantages Placement methods

(i) Salt concentration becomes more in soil solution.

(ii) Some times plants are destroyed in the beginning stage.

(iii) If anyhow excess amount of fertilizer is collected near the seed, seed germination is badly affected.

2. Application of fertilizers in liquid form

1. Starter solutions
2. Foliar Spray
3. Application with irrigation water
4. Direct use of fertilizers solution in soil
5. Direct use of liquid fertilizers
6. Soaking of seeds in fertilizers solution
7. Injection of fertilizers solution into plants.

1. **Starter solutions**: At the time of transplanting of vegetables starter solution of N: P: K in 1:1:1or 1:1:2 or 1:2:1 ratio is applied. Very less amount of fertilizers is mixed and it is used at the place of water to establish plants. Starter solutions supply plant nutrient in the beginning of plants growth.
2. **Foliar Spray:** Dilute solution of the fertilizer is sprayed on plants in standing crop. One or many nutrients are sprayed together. plants absorb nutrients very soon thus the effect of foliar application in more than soil application Foliar spray is done by sprayer.

Foliar spray is useful in following ways-

(i) It is useful in dry farming and in crops sown late

(ii) Plant density is more in crop.

(iii) The effect of fertilizer is necessary at once.

(iv) When broadcasting of fertilizers is not possible.

(v) For spraying insecticides and pesticides.

Advantages of foliar spray

(i) Fewer amounts of fertilizers are required.

(ii) Fixation of nutrients does not occur in soil.

(iii) Uniform distribution of fertilizer occurs in the whole field.

(iv) Foliar spray of micronutrient is very useful because soil application causes toxicity.

Disadvantages of foliar spray

(i) Labour charge is more in spraying.

(ii) Spraying of high concentrated solution leaves is scorched.

(iii) It is useful when less amount of nutrient is available to plant.

(iv) On spraying of urea solution, some time the maximum percentage of biuret reaches in leaves which cause toxicity.

What precautions to be taken at the time of spraying.

1. Spraying should be done at proper time of plant growth stages.
2. Concentration of fertilizer should be made according to the type and stage of crop and types of sprayer.
3. Spraying should be done in the morning after drying dew. Spraying should be done after 12.00 P.M.
4. For uniform spray fine, nozzle should be used in sprayer.
5. Spraying should not be done on the possibility of raining or fast wind blowing.
6. Urea used for spray should not contain biuret more than 1.0 %.

Crop	Low volume sprayer (100-500 Lha^{-1})	High volume Sprayer (400-800 L ha^{-1})
Paddy	10-20 %	2-4%
Maize	10-20 %	2-4%
Wheat	10-30 %	3-5%
Jowar & Bajra	10-20 %	3-6%
Cotton	10-15 %	1-2%
Sugarcane	15-30 %	4-6%
Jute	10-15 %	1-2%
Oily crop	10-18 %	1-2%
Vegetables	5-10 %	0.5-1.0%
Flower plants	1-2 %	0.5%
Fruit plants	3-5 %	0.5-1.0%

Source: A text book of soil science & fertilizer manure-Dr. Lal Singh Nirankari

Spray of micronutrient: Foliar application of micronutrient is also done according to the need of crops. The concentration of solution is given below.

Elements	Conc. of compound in soln. (%)
Iron	0.4 $FeSO_4$ + 0.20 $Ca(OH)_2$
Manganese	0.60$MnSO_4$ + 0.30 $Ca(OH)_2$
Copper	0.10 $CuSO_4$ + 0.05 $Ca(OH)_2$
Zinc	0.50 $ZnSO_4$ + 0.25 $Ca(OH)_2$
Boron	0.20 Borax
Molybdenum	0.05 sodium molybdate

3. **Application with irrigation water:** Soluble NPK fertilizers are added in irrigation channel from which nutrients reach with irrigation water in whole field. By this method fertilizer is uniformly distributed in whole filed.
4. **Direct use of fertilizer solution in soil:** The use of fertilizers in our country is limited. Some solution are directly used by instrument and filed is always kept moist.
5. **Direct use of liquid fertilizers:** In our country direct use of liquid fertilizer is not prevalent but this method is very common in foreign country. For use of liquid fertilizers the field should be well ploughed and it should contain sufficient moisture. Ammonia becomes fixed on soil colloids at once by cation exchange capacity phenomenon.
6. **Soaking of seeds in fertilizers solution:** Before sowing of the seeds are dipped in dilute solution of fertilizers for a definite period. The seeds absorb the fertilizer solution which is utilized from seed germination to the beginning of plants growth. Some times seed germination is adversely affected by this method.
7. **Injection of fertilizer solution into plants:** Dilute solution of fertilizers

is injected into plants and added deep in soil by special instruments. Generally micronutrients solutions are injected into xylem of plants which are tranlocated in all parts of plant. In soil both macro and micronutrients are injected by specially instruments.

Fertilizer use efficiency: Chemical fertilizers and organic manures are applied to the soil to meet the additional requirements of crops and thereby to sustain the yield. Fertilizers are stoichiometric low molecular weight compounds, used either singly or in mixtures. By comparison, manures are mixtures of many non stoichiometric substances, usually bulky, carbonaceous and of high molecular weight. Both the categories have to be converted into forms suitable for crop uptake. This conversion is carried out in soil with the help of microorganisms. Urea is converted into NH_4^+ ion by enzyme, *urease* and then into nitrogen oxides and finally to NO_3^-. A part of nitrogen escapes as N_2O as well as molecules nitrogen gas by the process of denitrification.

Phosphatic fertilizers should be capable of releasing H_2PO^{4-} or HPO_4^{2-} ions for absorption by the crops. Nitrate, phosphate and potassium, if present as fertilizers are taken up directly by the crops. Manures, on the other hand, have to be mineralized by appropriate microorganism and then converted into forms suitable for absorption. This is a slow process and hence; manures ate not able to meet the immediate needs of crops. The most pertinent question is what percentage of an applied fertilizers nutrient is utilized by the crop /crops/ cropping system. This is a measure of what has been termed fertilizer use efficiency. The crop being the best judge it can be measured either by the percentage of added fertilizer nutrient recovered by the harvested portion of a crop or kilograms of economic produce per kilogram of nutrients applied. Of the major plants nutrients, nitrogen and phosphorus have received the greatest attention. Interestingly, the problems in regard to one are just opposite to those of the other. Thus, leaching and volatilization loss are the problems of nitrogen, fertilizers whereas fixation, immobility and transformation are the problems related to phosphorus.

Nitrogen use efficiency: It is mainly determined by various kinds of losses in the field viz. volatilization in the form of ammonia, denitrification, leaching and runoff. The nitrogenous fertilizers are mostly in amide and ammoniacal forms converted into nitrate form, it become very much susceptible to loss by leaching with irrigation and rain water. Application of nitrogen fertilizers in splits doses or as top dressing, therefore, increases the efficiency of nitrogen use.

The NH_4^+ions in ammonical fertilizers are absorbed by the soil clay. They are be utilized directly by certain crops, e.g. rice, else they are transformed into nitrate by microbes and taken in this forms. The higher the clay content of the soil and its exchange capacity, the better is the nitrogen use efficiency if the rate of nitrification is more than the rate of uptake by the crops, the surplus is lost through the leaching.

Urea the cheapest nitrogen source is transformed in the soil as follows by microbes

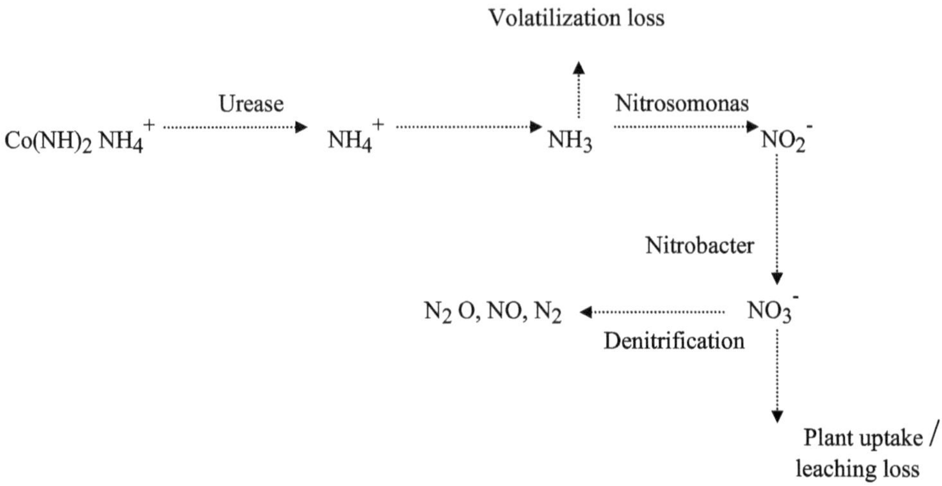

The main factors that determine nutrient uptake by crop apart from the nature of the roots system which varies from crop to crop are water texture and pH of soil, management practices and agro- climate conditions. The recovery of applied nitrogen as recorded from field experiments conducted in India has been found to vary from 28 to 34 % for rice and 40-60 % for other crops. The low recovery is the result of losses caused by volatilization, denitrification, leaching and run off. Nitrogen use efficiency depends not only on the source rate time am methods of fertilizers application but also on a large number of soil and crop management factors such as land preparation, choice of variety, timely sowing optimum plant population timely weed control proper water management plant protection and balanced supply of essential plant nutrients.

To increase the nitrogen use efficiency, the following practices are generally useful. Split application of nitrogen is a widely accepted practice for all crops.The basic approach is to apply nitrogen in two or more installments to coincide with peak period of nitrogen requirement of the crops. Some of the techniques that are likely to increase the nitrogen use efficiency are subsurface application. Pallets made by mixing nitrogenous fertilizers (particularly urea) with soil. prior incubation of urea with moist soil (1:6) for two or three days, resulting in the adsorption by soil clays of ammonia formed by the hydrolysis of urea.

Attempts have been made to increase efficiency through the use of slow release nitrogenous fertilizers. They are manly : (1) urea–aldehyde condensation product e.g. urea form (urea formaldehyde compounds) Oxamide and isobutylidene diurea (IBDU) and (2) urea coated with sulphur (SCU) and lac (LCU) and Neem and cakes. The coating is meant to form a temporary barrier between the urea granules and soil or soil water thus, reducing the rate of urea hydrolysis.

The other approach has been to blend nitrogen fertilizers with a *nitrification inhibitors* which retards the rate of transformation of NH_4^+ to NO_3^-, thereby making the fertilizers nitrogen gradually available to the plants. The widely used nitrification inhibitors materials *viz.* N- serve or Nitrapyrin (2-chloro-6 trichloro- methyl pyridine) AM (2-amino -4-chloro-6-methyl pyrimidine), and ST (2-sulphanilamide) have been found to increase nitrogen use efficiency. The presence of pyridine in the first two inhibit the action of nitrifiers. Such materials as Neem and karanj cakes and citrusllus colosynthis also inhibit nitrification. In the context of what has been stated earlier a judicious mix of manures and fertilizers also acts as a slow release system. Placement of urea super granules, made up of ordinary granules prilled urea and urea briquettes below the soil surface has been found to increase to varying extents the nitrogen use efficiency with regard to rice crop.

Phosphorus use efficiency: In case of phosphorus fertilization fixation of phosphate is the main problem. Water soluble phosphatic fertilizers soon after application to the soil react preferable with iron and aluminium to form sparingly soluble products called the *initial phosphate reaction products*. Calcium and may be magnesium carbonates, pH and water status of the soil control the nature of the reaction products. The efficiency of phosphatic fertilizers depends primarily upon the release of phosphorus form the products rather than the fertilizers per se. The reaction products vary in their ability to release phosphorus. One of the measures taken to improve phosphorus use efficiency is to have minimum contact of the fertilizers with the soil to restrict phosphate fixation. Another measure to increase it is to raise the pH, preferable of acid soil by liming. Liming deep placement and combined use of super phosphate and organic manures are known to increase the phosphorus use efficiency. Surface broadcast followed by maxing during Puddling has been found to give the highest phosphorus use efficiency for rice, being ever batter than placement at various depths (up to 0.15 m) and below the rows. Phosphate placement is definitely more beneficial than furrows or by drilling just below the seeds proved quite efficient for wheat.

Recovery of fertilizer phosphorus by crop is a single season is very low. Due attention should be given to the studies of utilizing the residual effect of the fertilizer. In other words, it would be advisable to adopt such a rotation of crops as would assure maximum utilization of phosphorus as direct as well as residual effect. To cite an example, the yield response percentage and recovery of fertilizer phosphorus by rice and gram crop are batter when super phosphate is applied to gram crop and its residual amount is allowed to be utilized by the succeeding rice crop than when it is applied directly to rice crop. In long term experiments at Puna Kanpur and Varanasi (UP) with rotation of follow-wheat–maize showed response to cumulative and direct application of phosphate at both the locations.

It has been observed in India that application of phosphatic fertilizers with organic manures is beneficial. This increases the crop response and maximizes phosphorus fixation. Soil organisms immobilize phosphate from labile and may remove as much phosphorus as the crop. There is a critical concentration of about 0.2% phosphorus in the organic matter being decomposed above and below which there will be net mineralization and immobilization. This should be kept in mind while applying organic matter to enhance the phosphate use. Experimental evidences show that there is better utilization of phosphorus on acid soils when mixed with farmyard manure. Long term experiments in India with sun hemp (*Crotolaria juncea*) have shown that an application of phosphate before ploughing is an excellent method for the improvement of soil fertility. The increased yield of green matter means more organic matter and nitrogen for the soils in both of which the soils of India are so much deficient.

Because of fixation, crops may not use more than 10% of phosphorus in the fertilizers applied broadcast and incorporated to the soil. However, up to 30% of higher may be used when applied as concentrated band along the plant row, while mixing with the soil increase fixation, localized placement of banding allows the phosphatic fertilizers to react with only a much smaller portion of the soil in the immediate vicinity of the band. Thus, more phosphate remains in an available form for the crop use. For banding water soluble phosphates are preferable. The greater the phosphate fixing capacity of the soil the greater will be the advantage of banding or of using large granules It may be noted that clay soils have greater phosphate fixing capacity than sandy soils.

Compared with water soluble form, the chance of fixation of citrate soluble form is less, hence such phosphatic fertilizers having high citrate soluble forms may be broadcast. Ground rock phosphate which is neither water nor citrate soluble should preferably be applied to acid soils, taking care to thoroughly mix it with the soil. This ensures reaction with the soil acids which bring phosphorus into available form. In that case, greater number of plant roots would come in contact with the soil particles and solubilize the rock phosphate better.

Organic manures, green manures grown with biological nitrogen fixating inoculants as well as organic wastes (not containing any toxic material)may be applied to the field prior to cropping. This steps increases the contents of NPK and other micronutrients of the soil, thereby increasing the fertilities use efficiency in general.

4

Diagnostic Techniques for Analysis

Diagnostic techniques: For obtaining maximum crop yields with maximum benefits to the cultivators, it is most essential that the crop plants should be fed properly with all nutrients. Soils deficient in particular nutrients must be supplied with fertilizers containing those plants nutrients. Thus, it is important to know which plant nutrients are lacking in a soil. Simple and elaborate tests have been developed by the agricultural scientists to estimate the nutritional requirements of soils and crops. These methods are known as diagnostic techniques. The wide variety of diagnostic techniques used so far can be broadly grouped into-

(i) Biological methods

(ii) Visual symptoms of nutrient deficiency or toxicity

(iii) Plant analysis

(iv) Soil analysis

1. Biological Methods: Biological methods utilize plants as extracting agents for nutrients from soil. In one group i.e. Field trials, pot culture and neubauer seedling methods, higher plants are used whereas in the other group lower plants, such as fungi and bacteria are the test organism.

Field trials: The most direct way of diagnosing the nutrient status of a soil is field trial in which the effect of additional amounts of nutrient in the forms of fertilizers applied to the soil on the growth of the crops is measured. Fertilizer nutrients in graded doses are applied to statistically laid out plots in a fairy homogenous field. Yield is the main criteria for assessing the nutritional requirement of crops in that soil. Trials with graded doses of nutrients will determine the exact requirement.

Broad generalization can be made from the field trials regarding the kind and quantity of nutrients needed for optimum yield of crop or crops. But, the effect is very much location and situation specific and is dependant on climate conditions. But this is the only direct method. Field trial is expensive and time consuming and also the growth conditions can not be fully controlled. However, whatever diagnostic techniques for nutrient assessment and fertilizer requirement may be developed field trial is the ultimate test to which the findings of any other proposed method must be subjected. If a yield prediction fails to be confirmed by field trial such a method is not acceptable.

Pot culture methods: This method was first initiated by Mitscherlich. The principle is the same as in the field trials except that plants are grown in a small volume of soils in plots or other container under controlled conditions e.g. in a green house. Comparative growth or yield of crops and uptake of plant nutrients from the treated and untreated portions of soil are usually taken as indices of nutrient deficiency. The limitations of plot culture methods are preferential roots growth, limited volume of soil and problems encountered in the packing of the soil in the pots for stimulated of natural field condition. In fact, chemical extractants for determining available nutrients are subjected to preliminary screening by pot culture methods.

Neubauer seedling methods: It is based on the principle of intensive uptake of nutrient elements by growing a large number of seedlings on a small quantity of soil. As a result, the soil is exhausted of the available nutrients. The entire plant in then analyzed for the uptake.

For example, 100 rye seedlings are grown for 14 to 17 days in Petri dishes (11x 7 cm) on a mixture of 100 g soil and 50 g of nutrient free quartz. The roots and tops are analyzed for K_2O and P_2O_5. Neubauer used this technique for phosphorus and potassium. This technique has been used for calcium and a few micronutrients. The Neubauer limits for a few crops are presented in given below.

Table: Neubauer limit values (mg /100 g oven dry soil)

	Crop	
	P_2O_5	K_2O
Barley	6	24
Oat	6	21
Potato	6	37
Rye	5	17
Sugar beet	6	25
Turnip	7	39
Wheat	5	20

The amounts of K_2O and P_2O_5 taken up by roots and tops less the amounts of K_2O and P_2O_5 in seed weights of the soil cultures represents the amount of root soluble K_2O and P_2O_5 in 100 g of oven dry soil and are usually designated as the "Neubauer number" expressed as mg per 100 g of dry soil. The neubauer limit values are assumed to be the minimum for satisfactory yields of the test crop.

Aspergillus niger method: *Aspergillus niger* a fungus is a biological indicator of available potassium phosphorus and magnesium in soils. In this method a small quantity of soil in liquid culture medium is inoculated with a suspension of A. Niger spores and incubated for four days. A thick pad of mycelium grows on the surface of the medium. The weight of dry mycelium and the content of K_2O absorbed by mycelium from oil were used by Mehlich to determine deficiency or sufficiency of potassium. The amount of potassium extracted in the mycelium compares favorably with the content of exchangeable potassium of the soil under investigation. The standards used by Mehlich are presented in table as given below.

Table: Standards for K status of soils by A. niger method

Weight of Four pads (g)	K_2O absorbed by A. Niger/ 100 g soil (Mg)	Degree of K deficiency
< 1.4	< 15	Very deficient; indicates need for K fertilization.
1.4 to 2.0	15 to 20	Moderately to slightly deficient; light application of K fertilizer may be beneficial;
>2.0	>20	Not deficient, no need for K fertilization

Soil plaque method: Winogradsky demonstrated that spontaneous colonies of *Azotobacter* grow luxuriantly on soil plaques if suitable energy materials are provided and also observed a close correlation between the limiting factor (say nutrient) for *Azotobacter* and those for growing plants. On this basis, Sackett and Stewart developed a soil plaque technique for study of mineral deficiency in soil. A small portion of soil with corn starch or sucrose is inoculated with *Azotobacter* and incubated. If the soil is deficient in potassium or phosphorus, the colonies of the organism will not develop. In the presence of either P or K or both, colonies develop and this indicates the soil deficiency in K or P both. The intensity of growth of colonies indicates the degree of deficiency. This technique has been successfully used to indicate availability of potassium and phosphorus.

Use of visual symptoms of nutrient deficiency or toxicity: Typical symptoms of deficiency develop in plants when the nutrient supply is below a certain critical level and those of toxicity above a certain level. It is common knowledge that when nitrogen is in short supply, plant show stunted growth and pale colour particularly the leaves. Application of nitrogenous fertilizers removes the malady. Thus, growing plants can be used an index of the nutrient level of soil or other growing medium several colour photograph have been published showing symptoms of deficiency or toxicity of different species of plants.

The practical applicability of this method of diagnosis is rather limited. The symptoms are very often vitiated by the interaction of the elements and also by the infestation of pests and pathogenic diseases. Even when the symptoms has been diagnosed properly it may be too late to remedy it and get a reasonably yield by applying the nutrient in question. The type of symptoms included by a nutrient deficiency varies from species to species and even from variety to variety. The exact nutrient causing the symptoms may be determined from plant analysis or by foliar application of solutions containing soluble compounds of the nutrient or different nutrients and observing the disappearance of the deficiency symptoms. The appearance of deficiency symptoms is an extreme limit of nutrient deficiency but even if the symptoms are not manifest diminution in the field of crop may occur. This condition has been termed as ***hidden hunger***. Notwithstanding this limitation, visual deficiency symptoms have been used successfully in correcting deficiency in perennial crops, particularly horticultural crops.

Plant analysis and soil analysis: A research worker in soil fertility is interested in finding out the nutrient uptake by crop at different stages. The stages at which nutrients uptake is to be studied will very with different crops and their varieties. One may also be interested in collecting a plant samples when some abnormality is observed in the plants and a nutrient deficiency or toxicity is suspected. In such cases no time should be lost in collecting the sample otherwise secondary reactions that often take place in the mineral content of plants will conceal the real cause of deficiency or toxicity it may also be mentioned hare that in such cases only the affected plants are to be sampled.

Area that can be represented by one sample: For research projects each plot receiving differentially fertilizer or some other treatment is a separate until and is to be sampled separately. For general advisory purpose one sample would be good enough for the entire field (not exceeding 5 acres). One should, however, take care that the sample comes from a field or a part there of having a uniform sand. If a field is patchy and differences in stand and plant growth are obvious, each area should be sampled separately.

Sampling techniques: The person collecting the sample should travel across the rows in the centre of the plot. In small plots every alternate row and in big fields every 5th to 50th row should be sampled. For small experimental plots (say about 100 sq.m.) about 10 plants would serve the purpose but for bigger fields about 25-50 plants should be collected. It is always advisable to collect a larger sample in the field and then to reduce it in the laboratory as it helps in generating a representative sample.

Tissue to be samples: For working out the plant nutrient uptake by crop at different stages entire plants are to be sampled. Different tissue (generally leaf and stem but any other part can be separately). If desired should then be separately (their dry weight must be recoded) and analyzed separately. At harvest grain and straw samples are collected and analyzed separately.

Soil pH: Soil pH may be defined as the logarithm of the reciprocal of the active hydrogen ions in gl^{-1} (gram per litre).Thus; it is the measurement of H+ ion activity. Such measurements are intensity measurements and differ from capacity measurements (such as exchangeable H^+) where absolute amounts are involved. It would be of much use to know the pCa, pMg and the pK (activities of Ca, Mg, and K, respectively) since these measurements are related to nutrients activities in the plant root environments. The pH of the soil water system is related to the ionic saturation of the soil colloidal complex with reference to the ratio of H and (Ca + Mg).

Measurement of soil pH is affected by several factors. Some of these are:

- Soil to water ratio employed
- Partial pressure of CO_2
- The salt content of the sample
- Drying of the soil samples and
- The amount of grinding given to the soil.

Soil pH is still mostly determined in soil–water suspension. The amount of water added varies from enough to bring the soil to saturation to ten times the weight of soil. Scholfield and Taylor (1955) have shown that a much reliable measurement of soil pH can be made in a 0.01 M $CaCl_2$ solution. Determining soil pH in 0.01 M $CaCl_2$ is quite useful and informative and can be used for determining lime potential. In several European countries soil pH is determined in 1M KCl solution.

Determination of soil pH

Reagent: 0.01M $CaCl_2$ solution (2.18 g. $CaCl_2.6H_2O$ or 1.46 g $CaCl_2$ $2H_2$per litre)

Procedure: Take 20 g soil in a 100 ml beaker add 50 ml of distilled water and measure pH on a pH meter using a glass and a calomel reference electrode.

Or

Suspend 25 g soil in 50 ml 0.01 M $CaCl_2$ solution and measure pH with a glass electrode and a calomel reference electrode on a pH mater.

Comments:

- Soil pH determined in 1M KCl or 0.01 M $CaCl_2$ is generally lesser (0.2 to 1.0 units) than that determined in water. This is because of K or

Ca ions in solution exchange with exchangeable H^+ on clay particles and thus more H^+ ions come in solution resulting in lower pH values.

- Lime potential as given by Schofield and Taylor (1955) is expressed as pH-/2p Ca. When soil pH is measured in a 0.01 M $CaCl_2$ solution, lime potential can be easily determined. Since, 1/2p Ca in 0.01 M $CaCl^2$ is 1.14 (Hesse, 1994). Thus, if pH of a soil as determined in a 0.01M $CaCl_2$ is 7.2, the value for lime potential will be 7.2- 1.14 = 6.06

Organic Matter in Soil: Organic matter in a mineral soil is regarded as an index of its fertility status. In India organic carbon content of soils is used for evaluating the nitrogen supplying capacity of soils.

Organic matter content of soils is determined by three sets of methods:

- Total organic matter determination by H_2O_2 oxidation and weight loss
- Organic C determination as CO_2 either by dry combination or by wet oxidation by chromic acid and
- Determination of oxidizable organic matter.

The weight loss methods are susceptible to errors caused by presence of salts having carbonate, water and hydroxyls since these influence the weight change. As regards the other methods, although chromic acid is used in both the wet oxidation and determination of oxidizable OM, the two methods differ in technique as well as in principle. As regards technique CO_2 is evolved in wet oxidation method and is estimated gravimetrically or titrimetrically, whereas the amount of chromic acid reduced is measured titrimetrically in the methods for oxidizable forms and in carbonates elemental C in the form of graphite, coal, charcoal or other highly resistant forms and the easily oxidizable organic C is involved and carbonate and elemental C are excluded. In wet oxidation method correction for carbonate content has to be made.

In the procedure described here, organic C is oxidized to CO_2 by potassium dichromate which in its turn gets reduced. The excess of potassium dichromate left is titrate against a standard ferrous sulphate solution and the exact amount of potassium dichromate required to oxidized the soil C is determined (Walkley and Black, 1934).

The oxidation of C can be expressed by the following equations:

$$4\ Cr^{6+} + 3\ C = 4\ Cr^{3+} + 3C^{4+}$$

$$2H_2Cr_2O_7 + 3\ C + 6\ H_2SO_4 = 2Cr_2\ (SO_4)_3 + 3\ CO_2 + 8\ H_2O$$

However, C is not the only element that is oxidized by $H_2Cr_2O_7$ but H^+ also reacts. Presence H^+ therefore increase the amount of dichromate required for oxidation. However, the C present in the carboxylic groups does not require dichromate to get oxidized (R-COOH= RH+ CO_2) and thus, lowers the amount of dichromate required for oxidation. It is generally accepted that the opposite effects of H^+ and O^- (in carboxylic groups) balance each other.

Walkley and Black Method:

Reagent:

- **1N potassium dichromate**: Dissolve 49.04 g potassium dichromate in water and dilute to one litre.
- **0.5 N Ferrous sulphate**: Dissolve 140 g of $FeSO_4.7H_2O$ in water, add 40 ml conc. H_2SO_4 cool and dilute to 1 litre. This reagent should be standardized every day before use.
- **0.25 N O-phenanthroline ferrous complex indicators**: Dissolve 1.485 g of O-phenanthroline molybdate and 0.695 g $FeSO_4.7H_2O$ in distilled water and dilute to 100 ml (also available directly as prepared solution) colour change pale blue to chocolate red. Or Diphenylamine indicator prepared by dissolving 0.5 g diphenylamine in a mixture of 100 ml conc. H_2SO_4 and 20 ml water. Colour change blue to green.

In the oxidation of Fe^{+2} to Fe^{+3} the redox potential changes from 0.94 to 1.40 and the end point of titration is at about 1.3 O-phenanthroline changes its colour pale blue to chocolate red at a redox potential of 1.14 and therefore suits ideally for this titration. On the other hand diphenylamine changes its colour at a redox potential of 0.76 and can be used only when phosphoric acid and sodium fluoride are used to complex Fe^{+2} by lowering the conc. of Fe+3 the end point is shifted to a redox potential of about 0.76 and diphenylamine can be used as in indicator.

- **Conc. sulphur acid sp.gr. 1.84, C.P.**
- **Conc. phosphoric acid 85% C.P.** (To be used only diphenylamine indicator)
- **Sodium fluoride (Na F)** (To be used only diphenylamine indicator)

Procedure: Take a suitable amount of soil (2 g for soils having less than 1% OM, 0.5 g for mountain soils and 0.05 g for peat soils) having passed a 80 mesh sieve in a 500 ml. Erlenmeyer flask and add 10 ml of 1 N $K_2Cr_2O_7$ solution. Swirl gently to mix the reagent with soil thoroughly. Now add slowly along the sides of the flask 20 ml of conc. H_2SO_4. Mix acid with soil by rotating the flask gently and leave the flask on an asbestos sheet for 30 minutes to allow the oxidation to proceed. Now diphenylamine is used as an indicator. Titrate the contents of flask with $FeSO_4$ solution; add $FeSO_4$ solution freely from the burette until the contents of the flask turn dull green; then start adding it by drops until colour changes to chocolate red. Carry out a blank determination. If more than 8 ml of dichromate are reduced the determination should be repeated with lesser amount of soil.

Calculation

$$\text{Organic carbon (\%)} = \frac{10\,(\text{Blank reading} - \text{Sample reading})}{\text{Blank reading} \times \text{Wt. of soil in g}} \times 0.003 \times 100$$

% Actual organic carbon in soil = OC estimated x 1.3 **

% O.M. in soil = Actual C% x 1.724***

(***The Van Bemmelen factor of 1.724 is used because OM contain 58% C. ** indicates based on 77% recovery)

Comments

In the oxidation of organic carbon the change in valence is form o to +4 (in CO_2). The equivalent of C in this reaction is 12/4= 3, thus, 1 ml of 1N $K_2Cr_2O_7$ oxidize 3 mg or 003 of C.

Total Nitrogen (N) in Plants: Total N in plants and soil samples is one of the most frequent determinations made in soil fertility investigations. This determination is needed in estimating fertility status of soils, working out an N balance sheet, studying relative efficiency of N sources and investigations on the effect of various agronomic practices on soil N etc. The two standard procedures for determining total N in soil or plant samples are:

Colorimetric Method: A colorimetric procedure can also be followed either manually or by using auto analyzer system for determination of NH_4-N in the plant digest (Baethgen and Alley, 1989).

1. *Kjeldhal Method:* It method involves digestion of the sample to convert the N compounds in the sample to the NH_4^+ form.

Principle: Nitro compounds formed by the reaction of salicylic acid with NO_3 in acid medium are reduced to corresponding amino compounds by heating the mixture with sodium thiosulphate and zinc dust. The main product of nitration is 5-nitrosalicylic acid and small amount of 3-nitrosalicylic acid. Drying of plants samples in oven eliminates interference of water in analysis as nitration of salicylic acid does not take place if more than a trace of water is present.

Reagents

- **Sulphuric salicylic acid mixture**: 1 g of salicylic acid to 30 ml of conc. H_2SO_4
- **Sodium thiosulphate**
- **Sulphate mixture**: 20 parts of K_2SO_4 + 1 parts catalyst mixture (20 parts $CuSO_4$ + 1 part selenium powder).

Procedure: Take 0.5 g dried plants tissue is transferred to a 100 ml Kjeldhal flask. And 20 ml of sulphuric salicylic acid mixture is added and swirled gently so as to bring the dry sample to come in contact with the reagents. It is allowed to stand overnight. Next day 5 g of sodium thiosulphate is added and heated gently for about 5 minutes. Care should be taken to avoid frothing. The contents are cooled; 10 g of sulphate mixture is added and digested on Kjeldhal apparatus at full heat. The digestion is continued for 1 hours after the solution is cleared. Bumping during the digestion can be avoided by addition of glass beads. When the digestion is complete, the digest is cooled and diluted.

Distillation: Ten ml of the digest is taken and transferred to vacuum jacket of micro-Kjeldhal distillation apparatus. In a conical flask, 10 ml of 4% boric acid solution is taken containing bromocresol green and methyl red indicator to which the condenser outlet of the flask is dipped after adding the aliquot the funnel of the apparatus is washed with 2-3 ml of deionized water and 10 ml of 40% NaOH solution is added.

Five ml aliquot is distilled to the flask containing 10 ml of boric acid. After completion of distillation, the boric acid is titrated against N/200 H_2SO_4. Blank should also be run and titration should be carried out to the same end point as that of sample. The Nitrogen content in plant sample is calculated as follows:

- Weight of sample = 0.5 g
- Normality of H_2SO_4 = N/200
- Volume of digestion = 10 ml
- Aliquot taken = 5 ml
- Titration value (TV) = Sample titration value – Blank titration value

N in 5% =

$$\frac{TV \times 0.00007 \times 100 \times 100}{0.5 \times 5} \quad (1 \text{ ml of N/10 } H_2SO_4 = 0.00014 \text{ g N})$$

Example TV= 3.6 ml

TV × 0.28 = 3.6 x 0.28 = 1.01% N

Note: Automated versions of Kjeldhal distillation have been developed and can be used

Functions of reagents

1. **Conc. H_2SO_4 acid:** Oxidation of soil C or plant C to CO_2.
2. **Copper sulphate:** Catalyst in oxidation of C by sulphuric acid.
3. **Sodium thiosulphate:** It reduces the NO_2 group on salicylic acid to $-NH_2$ group.
4. **Salicylic acid:** This is used when nitrates are to be included in the Kjeldhal–N analysis.
5. **Boric acid:** For trapping ammonia coming out from distillation.
6. **Std. Sulphuric acid:** For estimating ammonia trapped in boric acid.

Available Nitrogen in Soils: Assessment of soil's fertility status involves an estimation of its available nutrient status. About 95 % or more of soil N in mineral soil exits in organic form and its availability to plants depends much upon the microbial activity in soil. Although total soil N content of a mineral soil gives some idea as to its N supplying power, the practical value of a reliable method providing an index of the availability of soil N has long been appreciated. This exercise is commonly referred to as soil testing and it's used by farmers around the world to arrive at optimum fertilizer application.

Preparation of sample for analysis: Sample brought to the laboratory is spread out on an aluminium tray or plastic or thick brown paper. It is preferable to air dry soils at 20-25°C and 20 to 60% relative humidity (Jeckson, 1958) Samples may be mixed during drying to expose fresh surfaces. After air drying soil samples are crushed gently in pestle and mortar and sieve through a 2 mm sieve. Crushing should be continued until the soil retained on the sieve contains no aggregates. The material larger than 2 mm sieve is discarded. If the soil contains soft chalk or lime stone, the degree of grinding influence the content of $CaCO_3$. The pestle and mortal may be of wood, porcelain /stoneware (agate) or stainless steel. Sieve can be stainless steel or nylon.

Nearly all determinations are carried out on the fraction less than 2 mm. If less than 1 g of sample is required for a particular analysis(say, organic carbon)then 25-50g of 2 mm fraction is further ground with pestle and mortal to pass through 0.5 or 0.2 mm sieve. If soil sample size is large, it is again reduced by 'quartering technique' or using 'sample splitter' or 'riffle sample'. For certain analysis or experiments where field moist samples are required these must be transported in plastic bags immediately after collection. After rapid removal stones etc., these samples are rubbed through a wire mesh sieve with openings about 4-5 mm across and weighed immediately and dry weight determination (Dewis and Freitas, 1984).

Easily oxidizable organic carbon and mineralizable N are largely used as measures of available N in Indian soils. Organic C is estimated by the modified Walkley-Black method (Walkley-Black, 1934) and available N by the method of Subbiah and Asija, (1956). Some important chemical methods for estimating available N in soils are:-1. Total soil N content 2. Organic matter content 3. Nitrogen hydrolyzed by various regents e.g. H_2SO_4 NaOH, alkaline permanganate oxidation (Subbiah and Asija, (1956) and calcium hydroxide hydrolysis (Prasad, 1965). This methods involves distilling the soil with alkaline potassium permanganate solution where determining the ammonia liberated. This serves as an index of the available mineralizable N status of a soil and was therefore proposed as a soil test for N by Subbiah and Asija, (1956).

Reagents:

- 0.32% $KMnO_4$ solution: Dissolve 3.2 g $KMnO_4$ in distilled water, make up to 1 litre in volumetric flask and mix well.
- **2.5 % NaOH solution:** Dissolve 25g NaOH flakes in distilled water, make up to the volume to 1 litre.
- Liquid paraffin (extra pure)
- 0.02% N (N/10) standard sulphuric acid

- Boric acid-indicator solution: Dissolve 20 g boric acid (H_2BO_3) in about 700 ml of hot water. Transfer the cooled solution to a 1 litre volumetric flask containing 200 ml of ethanol and 20 ml of mixed indicator solution. After mixing the contents of the flask and add a little 0.05 N NaOH cautiously until the colour is reddish purple (red), make up the volume and mix the contents of the volumetric flask.
- **Mixed indicator**: Dissolve 0.07 g of methyl red and 0.1 g bromocresol green in 10 ml of 95 % ethanol.

Procedure: Place 20 g soil in 800 ml Kjeldhal flask, add 20 ml of water and swirl the Kjeldhal flask. Now add 1 ml of liquid paraffin and few glass beads to prevent frothing and bumping during distillation and finally add 100 ml each of 0.32% $KMnO_4$ and 2.5% NaOH solutions. Distill the contents in a Kjeldhal assemble at a steady rate and collect the liberate ammonia in a 250 ml Erlenmeyer flask containing 20 ml of boric acid solution. With the absorption of ammonia the pink colour of boric acid solution turns to green. Nearly 100 ml of distillate is to be collected in about 30 minutes. Titrate the contents with 0.02 N H_2SO_4 up to end point (Pink colour).

Calculation

Mineralization N kg/ha = R × 31.36;where R= volume of 0.02 N H_2SO_4 required for titration.

Comments and suggestion

Limit for low, medium and high available N (kg/ha) for Indian soil are below 280, 280-560 above 560 respectively (Subba Rao, 1995).

- Prepare 2.5% NaOH solution fresh as NaOH absorb CO_2 and the strength alkali come down. Avoid blockade of mouth of the reagent bottle by the rubber stopper in place of glass stopper.
- Check leakage in the distillation until before use.
- Stem the apparatus with distilled water before use.

Available Phosphorus in Soils: Contrary to soil- N, the major part of soil –P is mineral in nature and its availability to depend much upon soil reaction. Almost all the methods of estimating available P in soils are chemical. In soils phosphorus exist in the form of various type of orthophosphates. A very small fraction of these is available to plants at a given time. Available phosphorus content of soil consists mainly of Ca Al-P and Fe-P. In the neutral and alkaline soil particularly Ca-P is the dominant fraction. Organic–P fraction is also in considerable amount, but is usually not included in the determination of available Phosphorus. A large number of extracting reagents from Dyer's 1% citric acid to some of the multi-nutrient extractants, buffer solutions, acids and chelating agents have been suggested for available P determination from time to time. However, no single extractant appears to be suitable for all type of soils. Two types of extraction methods are more popularly adopted. Under acidic soil conditions, Bray's P-1or Bray No.1 which involves soil extraction with a solution containing 0.03% N NH_4F and 0.025% N HCl is widely followed.

The fluoride complexes Al and Fe in soil thus, releasing some bound P besides the easily acid soluble P largely Ca-P. The other and most widely used extractant is the 0.5M $NaHCO_3$ solution at pH 8.5. The reagent is most suitable for neutral to alkaline soils and is designated to control the ionic activity of calcium through solubility product of $CaCO_3$ thus extracting the most reactive forms of P from Al, Fe and Ca-phosphate. Phosphorus in the extract can be determined using suitable method of colour development and measuring the colour intensity at an appropriate wavelength.

Olsen's procedure is most commonly used in India and is described below.

Olsen's method (Olsen *et.al* 1954)

Instrument

1. Colorimeter or spectrophotometer
2. Mechanical shaker

Reagents

- **0.5 M $NaHCO_3$:** Dissolve 42 g of P-free sodium bicarbonate in about 500 mL of hot water and dilute to one litre. Adjust the pH to 8.5 using dilute NaOH solution or dil. HCl.
- **Activated charcoal:** Wash pure activated charcoal or commercially available Dargo G-60 with acid to make P-free, even if having traces of P.
- **1.5% ammonium molybdate solution**: Dissolve 15 g of AR grade ammonium molybdate in 300 mL of warm water cool and add exact 350 mL of 10 N HCl. Make up the volume to one litre.
- **40% $SnCl_2$ solution:** Weigh 10 g stannous chloride in a 10 ml glass beaker. Add 25 mL of conc. HCl and dissolve by heating. Cool and store in an amber coloured bottle in dark, after adding a small piece of Zn metal to prevent oxidation. From this prepare a dilute $SnCl_2$ solution immediately before use.
- **100 mg P/L solution:** weigh 0.439 g of AR grade KH_2PO_4 dried in oven at 60°C for 1 hour in a one litre beaker add about 500 mL of distilled water and dissolve. Add 25 mL of approximately 7N H_2SO_4 and make the volume to one litre.
- **2 mg P /L solution:** Dilute a suitable volume of 100 mg P /L solution by 50 times to get 2 mg P per litre solution.

Procedure: Weigh 2.5 g of soil sample in 100 mL conical flask. Add a pinch of Darco G-60 and 50 mL of Olsen's reagent (0.5 M $NaHCO_3$). Shake for 30 minute on a mechanical shaker. Filter through Whatman No. 1 filter paper. Transfer 5 mL of clear and colorless filter into a 25 mL volumetric flask. Gradually add 5 mL of ammonium molybdate solution containing 400 mL of 10 N HCl per liter. Shake slowly and carefully to drive out the CO_2 evolved. When frothing completely ceases add distilled water, washing down the sides to bring the volume to about 22 mL. Add 1 mL of freshly diluted $SnCl_2$ solution shake a little and make the volume to 25 mL. Read the blue colour intensity at 660 nm (red filter).Run a blank without soil under identical manner.

Calculation

$$\text{Available P (kg/ha)} = \frac{Q \times V \times 2.24 \times 106}{A \times S \times 106} = \frac{Q \times V \times 2.24}{A \times S}$$

Where, Q = quantity of P in µg read on X-axis against a sample reading
V = Volume of extracting reagent used (mL)
A = volume of aliquot used for colour development (mL) and
S = weight of soil sample taken (g)

Thus, Olsen's P (kg/ha) = Q × 8.96

Preparation of standard curve for P

1. Take a series of 25 mL volumetric flask
2. Pipette out 0, 0.5, 1.0, 1.5, 2.0 and 2.5 mL of 2 mg P/L solution.
3. Add 5 mL of the Olsen's extractant
4. Add ammonium molybdate as for Olsen's method and proceed to develop blue colour.
5. measure blue colour intensity and draw a standard curve by plotting concentrations of P in µg against colorimeter readings
6. If a straight line is obtained, find out a factor for each reading.

Note:

- All glassware to be used for P determination should be cleaned with chromic acid followed by thorough washing to minimize contraindication.
- In case of delay in measuring colour intensity, cover the samples store properly and add $SnCl_2$ just before measurements.
- Colour intensity measurement should be carried out exactly 10 minutes after developing blue colour. In $SnCl_2$ methods colour starts fading soon after development.
- Ammonium molybdate forms yellow coloured molybdophosphate complex with P in soil extract.
- Stannous chloride reduces the yellow coloured molybdophosphate complex to blue colour due to reduction of Mo. The intensity of blue colour gives the amount of P in soil extract.

Available Potassium in Soils: The term available potassium (K) conventionally refers to exchangeable water soluble K. The exchangeable K constitutes the major potion of available K except in saline or saline-sodic soils. Available K or exchangeable K, along with Ca and Mg are usually determined in neutral normal ammonium acetate (N NH_4OAC) extract in soil. The extraction is carried out by shaking followed by filtration or centrifugation. The K is estimated by using a flame photometer and Ca and Mg either by DTPA- titration or by using atomic absorption spectrophotometer (AAS).

The ammonium acetate method given below is most widely used. The ammonium ions are very close in size to K^+ and replace the lattice efficiently. During the estimation, acetate burns cleanly and does not leave any residue on the burner of flame photometer.

Ammonium acetate methods of K determination (Hanway and Heidel, 1952).

Instrument

1. Flame photometer
2. Mechanical shaker
3. pH-meter

Reagents:

- **1 N ammonium acetate, pH 7.0**: Dissolve 154 g ammonium acetate (CH_3COONH_4) in water and dilute to 1.8 liters. Mix thoroughly. Adjust ph to 7.0 with dilute NH_4OH or HOAC as required and make to 2 liters.
- **Potassium Chloride (standard solution):** make a stock solution of 1000 ppm K by dissolving1.908 g of AR grade potassium chloride in distilled water and diluting up to 1 litre. Prepare 100 ppm standard by diluting 100 ml of 1000 ppm stock solution to 1 litre with the extracting solution.
- **Standard curve:** Pipette 0, 5, 10, 15 and 20 ml of 100 ppm solution into 100 ml volumetric flasks and bring the volume to mark with extracting solution. The solutions contain 0, 5, 10, 15 and 20 ppm K respectively.

Procedure: Weigh 5 g of soil sample in 100 mL conical flask. Add 25 mL of the neutral *1N* ammonium acetate solution and shake for 5 minutes. Filter through What man No.1 Filter paper. Measure K concentration in the filtrate using flame photometer

Calculation:

$$\text{Available K (kg/ha)} = C \times \frac{25}{5} \times \frac{10^6}{10^6} \times 22.4 = C \times 11.2 \text{ or ppm K} \times 11.2$$

Where, C stands for the concentration of potassium in the sample obtained on X-axis against the reading.

Note:

- The filtrate should be clear in order to avoid choking of capillarity tube of the flame photometer which occurs very frequently.
- During the operation of flame photometer a constant air pressure and steady flow of gas and air combination is absolutely necessary for precise estimation. An appropriate combination is achieved by fixing the air pressure should around 0.45 psi and then adjusting the gas supply so as to get a non-luminous blue flame.
- Fans and coolers should be switched off so that flame is not affected
- Ammonium bottle should be cooled before opening.
- Potassium standards should be prepared fresh after every 2-3 weeks.

Comment

The ratings of soils on the basis of exchangeable K are; Low: < 108; Medium: 108-280 and High: > 280 kg/ha.

Available Sulphur in Soils

Sulphur: For estimating plant available sulphur in soils heat-soluble sulphur and 0.15% $CaCl_2$ extractable S methods (Williams and Steinberg, 1959) are usually recommendation.

Heat –Soluble S

Inorganic sulphate-S (SO_4-S) and a portion of organically held (SO_4-S) are mobilized by special heat treatments and extractant with 1.0 % sodium chloride. Sulphate-s is converted into $BaSO_4$ suspension using $BaCl_2$ crystal and a special conditioning reagent. The resulting turbidity is determined by a spectrophotometer (Cottenie *et al.*, 1979).

Reagents

- **1.0% Sodium chloride (NaCl):** Dissolve 10 g of NaCl in 700 ml of distilled water and make to 1 litre with distilled water.
- Barium chloride dihydrate ($BaCl_2$. $2H_2O$ crystals, 20 to 30 meshes):
- **Conditioning reagent**: Dissolve 75 g NaCl in 275 ml water in a 500 ml volumetric flasks stir well add 30 ml conc. HCl, 100 ml of absolute ethanol and 50 ml glycerol (Rinse it) Make up the volume with distilled water.

  Standard sulphate solution: Dissolve 0.5434 g K_2SO_4 (AR) in 1 litre of distilled water. This solution is 100 ppm. Dilute 10 times to obtain 10 ppm solution.

Procedure: Weigh 5.0 g soil into a silica basin and add 20 ml distilled water. Place the basin on a gently boiling water bath and evaporate to dryness. Then, heat it in a hot air oven at 102°C for 60 minutes. After cooling transfer the soil to a 50 ml centrifuge tube and extract it with 33 ml of 1.0% NaCl. Twenty five ml of aliquot is pippetted into a silica basin and evaporate to dryness with 2 ml of 3% H_2O. This is done to remove interfering organic carbon.

The basin is then heated in a hot air oven at 102°C for 60 minutes, to ensure the removal of excess H_2O_2. After cooling the residue is taken up in 25 ml water transferred to a centrifuge tube and centrifuged to remove suspended matter. Sulphur is then determined by taking a suitable aliquot. Pipette 10 or 20 ml extract in 150 ml Erlenmeyer flask and add 20 or 10 ml H_2O thus, bringing the volume to 30 ml. Add 2.5 ml stabilizing solution and 0.2 to 0.3 g of $BaCl_2$ crystal by a spatula to the standards and extract. Shake the flasks for 1 minute each at constant rate. After 1 to 3 minutes measure the turbidity in a colorimeter using blue filter. The turbidity can be more accurately measured at 340 nm on a spectrophotometer. The turbidity remains constant during 3-10 min.

Calculation

$$\text{Available S in soil (mg/kg)} = \frac{R \times 100}{10 \times 20} = \frac{R}{2}$$

Where, R stand for the quantity of S in mg as obtained on X-axis against a reading.

Comments and suggestions:

- Critical levels of available S (kg/ha) for different crops are: ground nut and mustard 10; wheat and maize 13; sorghum 8;finger millet 7; soybean 14 and sunflower 18 (Tandon, 1991).
- Wrap the rubber stopper in thin polythene otherwise errors result from gradual oxidation of s compounds present in the stopper.
- Turbid metric method is not as sensitive as the methylene blue method.
- Information on the precision of this method is not available.

0.15% $CaCl_2$ extractable S Methods

In this methods S is extracted with 0.15% and the soluble sulphate estimated turbid metrically on a colorimeter using blue filter or spectrophotometer at 340 nm.

Regents

Calcium chloride dihydrate ($CaCl_2$. $2H_2O$) 0.15%: Dissolve 1.5 g of $CaCl_2$.$2H_2O$ in about 700 ml of water and make to 1 litre with distilled water, and other reagents mentioned above are the same as for heat soluble S method.

Procedure;

- Transfer 5 g soil into a 150 ml Erlenmeyer flask.
- Add 25 ml 0.15% $CaCl_2$ to it.
- Shake for 30 minutes on a reciprocating shaker (180+ oscillations / min.)

- Filter the suspension through Whatman no. 42 filter paper.
- Analyze 10 to 20 ml extract turbid metrically as described for heat soluble S.

Calculation

$$SO_4\text{- S in soil mg/kg} = \frac{\text{S in aliquot (mg/l)} \times \text{Vol. of extract (ml)}}{\text{Volume of aliquot (ml)} \times \text{oven dry soil (g)}}$$

Comments and suggestions

- Its extract are clear and normally doing not require pretreatment with H_2O_2 to remove organic matter.
- Ca prevents deflocculation in clay soils and makes filtering more convenient.
- In neutral to alkaline soils, 0.15% $CaCl_2$ may serve as good an extractant as $Ca(H_2PO_4)_2$.

Available of Micro-nutrients: Because of the extremely small quantities needed by plants, soil testing for micronutrients elements namely Zn, Cu, Fe, Mn, B, Cl and Mo has been of little concern. Only in rare cases, the necessity of assessing their availability is felt. Now the deficiency of a few of them (e.g. Zn, Fe, Mn, and B) has appeared in certain areas mainly due to the use of high analysis inorganic fertilizers, little or no use of organic manures and heavy uptake by high yielding crop cultivars.

Less frequent deficiencies of these nutrients as compared to those of primary and secondary nutrients have also led to much less work done on development of suitable soil test methods for micronutrients. Nevertheless, a number of methods mostly colorimetric have been widely used for majority of the micronutrients. Neutral ammonia acetate and chelating agents like EDTA have been used as extractant for Fe, Zn and Mn and the extracted amount determined calorimetrically.

Zinc determined by dithizinate methods (Shaw and Dean, 1952) has been very popular until AAS was available. For those laboratories where the AAS is not yet available, the alternate methods as described by Jackson (1962) are still employed. However, for much rapid and accurate analysis of available Zn, Cu, Fe and Mn, the DTPA method (Lindsay and Norvell, 1978) as given below is most widely used. Available soil B is generally extracted by hot water and Mo by ammonium oxalate methods. The B in the soil extract is determined by Azomethine-H and Mo by dithiol method.

DTPA-$CaCl_2$-TEA method

Principal: DTPA, chelating agents combine with free metal ions in solution and forms soluble complexes. Due to the reduced ionic activity in solution desorption take places, bringing in some more ions from solid phase. DTPA offers the most favourable combination of stability constants for the simultaneous complexing of Zn, Fe and Mn. Since, Fe and Zn deficiency is frequently experienced in calcareous soils, the methods is designed to avoid excessive dissolution of $CaCO_3$ with the release of occluded micronutrients which are normally not available to plants. This is achieved by i) the inclusion of soluble Ca^{++} and ii) buffering the reagent at pH 7.3 with TEA (which burns cleanly during flame atomization). When the extractant is added to soil, addition Ca^{++} and some Mg^{++} enter the solution. This is largely because the proportional TEA exchanges with these ions from the exchange sites and this lead to the increased ionic conc. of Ca^{++} in the solution, which in turn helps in suppressing the dissolution of $CaCO_3$. DTPA extractant has the ability to Chelates most other chelating agents; it applies a moderate stress to solubilize soil Fe at a pH where $CaCO_3$ is not dissolved. Suitability of this method has been proved through excellent relationships between the test values and plants utilizable nutrients under pot and field studies conducted the world over.

Determination of available zinc

Instruments

1. Atomic absorption spectrophotometer (AAS)
2. Mechanical shaker

Reagents

- **Dilute HCl:** AR grade HCl diluted 5 times with double distilled water.
- **DTPA extractant:** Dissolve 1.967 g of AR grade Diethyline-Triamine-Penta Acetic acid (DTPA) and 1.470 g of $CaCl_2.2H_2O$ in about 25 ml of double distilled (DDW) by adding 13.3 ml of triethanolamine (TEA), followed by 100 ml more of DDW. Transfer the solution to one litre volumetric flask giving 4 to 5 washings. Just before making up the volume, adjust pH to 7.3 with dilute HCl. This reagent has 0.005*M* DTPA, 0.1M TEA and 0.01*M* $CaCl_2.2H_2O$.
- **Standard stock solution A (1000 mg Zn /L) :** Weigh exactly 1.0 g of pure Zn metal and dissolve it in minimum volume (10 ml) of dil. HCl (1:1) and make the volume to one litre.

Standard Solution B: Dilute 5 ml of solution A to 100 ml to get solution B containing 50 mg/L

Standard working solutions: dilute 0.5, 1.0, 1.5, 2.0, 2.5 and 5.0 mL portions of solution B to 50 mL to get working standards containing 0.5, 1.0, 1.5, 2.0, 2.5, and 5.0 mg Zn/L. The working standards should be prepared in the medium of the extracting solution after every few days as it can not be preserved for long.

Procedure

1. Weigh 10 g of soil samples in 100 ml conical flask.
2. Add 20 ml of the DTPA extractant and shake for 2 hr on a mechanical shaker.
3. Filter through Whatman no.42 filter paper, discarding first few drops. For quick filtration Whatman no.1 filter paper can also be used if the filtrate is clear.
4. Use the filtrate for Zn measurement using AAS.
5. Feed the standard working solutions and prepare a standard curve by plotting AAS reading against Zn concentration.

Calculation

$$\text{Available (DTPA- extractable) Zn in soil (mg/kg)} = \frac{A \times 20}{10} = A \times 2$$

Where, 'A' stands for the Zn concentration in aliquot as read from X-axis of std curve against the sample reading.

Determination of available copper: Available copper can be determined in the DTPA extract similar to Zn using AAS. For this, the standard stock solution can be prepared as given below.

Accurately weight 1.0 g AR grade copper mental wire or turning and dissolve it in 50 ml of diluted HNO_3 (1:1 with DDW) and finally make the volume to one litre. This is solution 'A' containing 1000 mg Cu/L. Prepare solution 'B' containing 50 mg Cu /L by diluting appropriate volume of solution 'A'. Finally prepare working standards containing 0.25, 0.5, 1.0, 1.5, 2.0 and 2.5 mg Cu /L.

Determination of available iron: Iron in the DTPA extract can also be determined with the help of AAS exactly in the same manner as Zn and Cu described above. However the working standard solutions of Fe should prepared for higher concentrations as the DTPA extractable Fe content of soil is generally more than both Zn and Cu. Thus, the Fe standard may be prepared as given below:

Prepare standard stock solution (solution 'A') by dissolving exact 1.0 g of AR grade Fe metal in about 50 ml of 1:1 diluted HNO_3 and dilute the contents to one litre with DDW. Prepare solution 'B' by diluting 50 ml of standard solution 'A' to 500 ml to get 100 mg Fe/ L. Finally prepare working standard solution containing 1.0, 2.0, 3.0, 5.0 and 10.0 mg Fe / L by diluting appropriate volumes of solution 'B' with the medium of extraction (DTPA).

Determination of available Manganese: DTPA extractable Mn is also determined following the same technique as adopted for Zn, Cu and Fe. For this prepare the standard solution as follows:

Weigh 1.583 g of AR grade MnO2 or 1.0 g of pure Mn metal and dissolve it in 50 ml of dilute HNO_3. Make the volume to 1 litre with DDW to get solution 'A' having a Mn concentration of 1000 mg /L. From solution 'A' dilute 25 ml to 250 ml with DDW to get solution B having 100 mg Mn /L. Finally prepare working solutions of 0.5, 1.0, 2.0, 2.5, and 5.0 Mg MN /L concentration by diluting 0.5, 1.0, 2.0, 2.5, and 5.0 ml portions of solution 'B' to 100 ml.

Determination of available Boron: Boron in soil exits in organic and inorganic forms. The compounds of boron, which have high availability, are water soluble. Boron deficiency is encountered in highly calcareous and acid soils while toxicity in salt affected soils as well as soils irrigated with high 'B' water. Soil extraction with cold or hot water continues to be the most acceptable methods for determining the available B in soils. In soils extract, 'B' can be determined by Azomethine-H method (John *et. al.* 1975). The methods employs Azomethine-H as the reagent to form a stable coloured complex of H_3BO_3 at pH 5.1 in aqueous media, which retains proportional absorbance concentration properties for several hours independent of the presence of a wide variety of salts. The maximum absorbance occurs at 420 nm. This technique is rapid, reliable and more convenient to use than traditional procedures employing carmine curcumin or quinalizarin.

Instruments

I. Refrigerator
II. Colorimeter or spectrophotometer
III. Polypropylene test tubes of 10 ml capacity

Reagents

- **Azomethine-H reagent:** Dissolve 0.45 g of Azomethine-h in 100 ml of 1% L-ascorbic acid solution in polypropylene reagent bottle. Fresh reagent should be prepared weekly and stored in refrigerator.
- **Buffer solution**: Dissolve 250 g of ammonium acetate (NH_4OAc) and 15 g of ethylene diamine tetra acetic acid (EDTA disodium) in 400 ml of distilled water. Slowly add 125 ml of glacial acetic acid and mix.
- **0.01 *M* $CaCl_2$ solution:** Dissolve 1.47 g of $CaCl_2.2H_2O$ in distilled water and adjust the volume to 1000 ml.
- **Boron standard solution:** Dissolve 0.114 g of boric acid (H_3BO_3) in distilled water and adjust the volume to 1000 ml. Each litre of this solution contents 20 mg of B. dilute 0.5, 1, 2, 3, 4, 5, 10, 20, 30, 40, and 50 ml of stock solution to 100 ml with distilled water to have solution with concentrations of 0.1, 0.2, 0.4, 0.6, 0.8, 1.0, 2.0, 4.0, 6.0, 8.0 and 10 mg of B / L respectively.

Procedure:

1. Weigh 20 g of air dried and 20 mesh sieved soil sample in a 250 ml B free conical flask.

2. Add 40 ml of 0.01 M $CaCl_2$ solution.
3. Attach a water cooled reflux condenser.
4. Heat the flask until first sign of boiling and then reflux the contents for exactly 5 minutes.
5. Allow to cool without removing the condenser.
6. Transfer a 1 ml aliquot of blank, diluted B standard or sample solution into 10 or 15 ml polypropylene tube.
7. Add 2 ml of buffer and mix.
8. Add 2 ml of Azomethine-H reagent, mix and after 30 minutes read the absorbance at 420 nm.
9. Refer these readings to that of standard curve prepared with 0, 0.1 0.2 0.4 0.6 0.8 1.0 2.0 4.0 6.0 8.0 and 10.0 mg of B / L solutions for converting to B conc. in soil extract.

Determination of available Molybdenum: Amongst the essential nutrients for plants, Mo is required in the smallest quantity. It is used by plants in the form of $MoO_4^=$ ions and its availability, unlike other elements is high in the alkaline pH range. As it is required in very small amount tests for availability of Mo are rarely carried out particularly in alkaline soils. Ammonium oxalate method is, however, the most popular for extraction of available Mo from soil. The extraction is advantageous as it is easy to prepare does not allow a significant change in pH during shaking due to buffering action and molybdenum forms stable complex with oxalic acid. The extraction is based on the replacement of $MoO_4^=$ ions from exchange sites which is irreversible due to formation of strong Mo-Oxalic acid complexes.

Instruments

I. Refrigerator
II. Colorimeter or spectrophotometer
III. Hot plate
IV. Water bath

Reagents

- **50% Potassium iodide solution:** Dissolve 50 g in 100 ml of double distilled water (DDW).
- **50% Ascorbic acid solution:** Dissolve 50 g in 100 ml of DDW.
- **10% Sodium hydroxide solution:** Dissolve 10 g of NaOH in 100 ml of DDW.
- **10% Thiourea solution:** dissolve 10 g in 100 ml of DDW and filter. Prepare fresh before use.
- **Toluene-3, 4-dithiol solution (Commonly called dithiol):** Weigh 1.0 g of AR grade melted dithiol (51°C) in a 250 ml glass beaker. Add 100 ml of the 10 % NaOH solution and warm the contents up to 51°C with frequent stirring for 15 minutes. Add 1.8 ml of thioglycolic acid and store in a refrigerator.
- **10% Tartaric acid**: Dissolve 10 g in 100 ml of DDW.

- Iso-amyl acetate
- Ethyl alcohol
- **Ferrous ammonium sulphate solution:** Dissolve 63 g of the salt in about 500 ml of DDW and then make the volume to one litre.
- **Extracting reagent:** Dissolve 24.9 g of AR grade ammonium oxalate and 12.6 g oxalic acid in water and make the volume to one litre. Adjust the pH at 3.3 with dilute HCl / dilute ammonia solution.
- **Standard stock solution (100 mg Mo /L):** Dissolve o.150 g of AR grade MoO_3 in 100 ml of 0.1 m NaOH make slightly acidic with dil.HCl and make the volume to 1 litre.
- **Working standard solution (1 mg Mo /L):** Dilute 10 ml of the stock solution to 1 L.

Procedure: Weigh 25g of air dried and 20 mesh sieved soil sample in a 500 ml conical flask. Add 25 ml of the extracting solution (1: 10 ratio) and shaker for ten hr. Filter through Whatman No. 40 filter paper. Collect 200 ml of the clear filtrate in a 250 ml glass beaker and evaporate to dryness on a water bath. Heat the contents in the beaker at 500°C in a furnace for 5 hrs. to destroy organic matter and oxalates. Keep overnight. Digest the contents with 5 ml of HNO_3-$HClO_4$ mixture (4:1) then with 10 ml of 4 N H_2SO_4 and H_2O, each time bringing to dryness. Add 10 ml of 0.1N HCl and filter. Wash the filter paper with separate 10 ml of 0.1N HCl and DDW until the volume of the filtrate is 100 ml, run a blank side by side or without soil

Colour development

1. Take 50 ml of the filtrate in 250 ml separatory funnels and add 0.25ml of ferrous ammonium sulphate solution and 20 ml of DDW and shake vigorously.
2. Add excess of KI solution and clear the liberated iodine by adding ascorbic acid drop by drop while shaking vigorously.
3. Add one ml of tartaric acid and 2 ml of Thiourea solution and shake vigorously.
4. Add 5 drops of dithiol solution and allow the mixture to stand for 30 minutes.
5. Add 10 ml of Iso-amyl acetate and separate out the contents (green colour) in colorimeter tubes/cuvettes.
6. Read the colour intensity at 680 nm (Red filter).

Preparation of standard curve for Mo.

7. Measure 0, 2, 5, 10, 15 and 20 ml of the working standard Mo solution containing 1 mg Mo /L in series of 250 ml separatory funnels.
8. Proceed for colour development as described above for sample aliquots.
9. Read the colour intensity and prepare the standard curve by plotting Mo Conc. against reading.

Calculation:

$$\text{Available Mo in soil (mg / g)} = A \times \frac{250}{200} \times \frac{1}{25} = \frac{A}{20}$$

Where, 'A' stands for the Mo concentration in aliquot as read from X-axis of std curve against the sample reading.

Methods of Reclamation of Problems soils: Some soil related problems render them unprofitable for cultivation. Even if the total elemental content in such soil is quite high these may not be fertile because of the poor soil health. The term soil health refers among other things to the often encountered soil problems like soil alkalinity/ sodicity, salinity and acidity.

Nutrient availability in such soils is severely affected due to adverse physical condition. Salts affected soils (saline, saline-alkaline and alkali) develop in arid and semi arid regions while acid soils are found in high rain fall areas where a considerable loss of alkaline earth metal cations takes place by ways of leaching. Poor drainage is a common feature of most saline soils and high salinity level of irrigation water further aggravates the soil salinity

As per the USDA classification the problem soils can be identified using the following **(Table-1)** characteristics determined in saturation paste of soil.

Characteristics	Saline soil	Alkali soil	Saline –Alkali soil
pH	<8.5	>8.5	<8.5
EC (dS/m)	>4.0	<4.0	>4.0
Exchangeable Na %	<15.0	>15.0	>15.0

Soil alkalinity is generally evidenced by sufficiently high pH to turn phenolphthalein to pink or red (pH 8.3 to 10.0 or 11). This pH range is often characterized by more than 15 % of exchange saturation with Na^+ ions. Poor permeability of the soil (Les than 1 cm /h) frequently develops with such high Na-saturation when salinity is low.

Determination of gypsum requirement of soil: Gypsum requirement of alkali soil can be determined by treating the soil with known amount of excess saturation gypsum solution and then estimating the unreacted or unutilized amount by versenate titration method as suggested by Schoonover (1952).Though Ca can be estimated any other methods also but the versene titration is more suitable.

Instrumental: Mechanical shaker

Reagents

- **Saturated gypsum solution:** Add 5 g of chemically pure $CaSO_4.2H_2O$ to one litre of distilled water. Shake vigorously for 10 minutes using a mechanical shaker and filter through Whatman No. 1 filter paper.
- **0.01 N $CaCl_2$ solutions:** Dissolve exactly 0.5 g of AR grade $CaCO_3$ powder in about 10 ml of 1:3 diluted HCl. When completed dissolved transfer to 1 litre volumetric flask and dilute to the mark with distilled water. $CaCl_2$ salt being highly hygroscopic should not be used.

- **0.01N Versenate solution:** Dissolve 2.0 g of pure EDTA- disodium salt and 0.05 g of $MgCl_2$ in about 50 ml of water and dilute to 1 litre. Titrate a portion of this against 0.01 N $CaCl_2$ solutions to standardize.
- **Eriochrome Black T indicator:** Dissolve 0.5 g of EBT dye and 4.5 g of hydroxylamine hydrochloride in 100 ml of 95 % ethanol. Store in a stoppered bottle or flask.
- **Ammonium hydroxide –ammonium chloride buffer:** Dissolve 67.5 g of pure ammonium chloride in 570 ml of conc. ammonia solution and dilute to 1 litre. Adjust the pH at 10 using dil. HCl or dil. NH_4OH.

Procedure: Weigh 5 g of air dry soil in 250 ml conical flask. Add 100 ml of the saturated gypsum solution. Firmly put a rubber stopper and shake for 5 minutes. Filter the contents through Whatman No. 1 filter paper. Entire quantity needs not be filtered. Transfer 5 ml aliquot of the clear filtrate into a 100 or 150 ml porcelain dish. Add 1 ml of the ammonium hydroxide-ammonium chloride buffer solution and 2 to 3 drops of Eriochrome black T indicator. Take 0.01N Versenate solution in a 50 ml burette and titrate the contents in the dish until the wire red colour starts changing to sky blue. Run a blank using 5 ml of the saturated gypsum solution in place of sample aliquot.

Calculation: Ca or Ca + Mg (me/L) in the aliquot = 2 V,

Where V stands for volume of versenate solution used

Since, 1 litre extract = 50 g soil (5 g soil to 100 mL).Ca retained (or Ca requirement) in me / 100 g soil.= [2V for added gypsum solution -2 V for filtrate] x 2

Gypsum requirement of soil in tonne per hectare up to 30 cm depth = A x 3.852

Apply correction pending on purity of gypsum.

Determination of lime requirement or Exchangeable –H in soils: Principle: In highly acidic soil having pH less than 6.0, nutrient availability is affected adversely. While availability of some elements (like B, Mo, etc.) decrease with fall in pH that of Fe, Mn, Al and many other heavy metals increase to the extent that these become toxic in certain cases. Reclamation of acid soils by liming is usually practiced for keeping the soil pH in the optimum range essential for making nutrient available in appropriate quantities. The amount of lime or liming material required depends on the nature of the soil. Through liming the exchangeable H of soil is called "Lime requirement of soil". The value of lime requirement depends on the extent to which the exchangeable H is to be neutralized as different crops demand variable soil pH for optimum growth and response. On liming Ca^{++} and Mg^{++} replace either Ca $(OH)_2$ or $CaCO_3$ on the colloidal surface. The liming material (CaO or Ca $(OH)_2$ or $CaCO_3$) react with CO_2 and water to form Ca $(HCO_3)_2$. By way of replacement of H^+ and Al^{+++} ions the Ca ions raise the percentage base saturation which leads to a corresponding rise in soil pH. The widely used method for determination of the lime requirement of soil (pH < 6.0 is that given by Shoemaker *et. al* (1961) which is described below.

Instrumental: pH meter

Reagents:

- **Extractable pH buffer solution:** Dissolve 1.8 g of nitro phenol 2.5 ml of triethanolamine, 3.0 g of potassium chromate, 2.0 g of calcium acetate and 53.1 g of calcium chloride dihydrate in one litre of distilled water and adjust the pH to 7.5 using dilute NaOH.
- **pH buffer solution:** As described under pH determination.(pH 4.0,7.0 and 9.2.)

Procedure: Weigh 5 g of air dry soil sample in 50 ml beaker. Add 5 ml of distilled water and 10 ml of the extractant buffer solution. Stir continuous for 10 minutes or intermittently for 20 minutes with glass rod.

Table 2: Lime required bringing the soil to an indicated pH using SMP Buffer suspension pH

Soil –Buffer	Agricultural ground lime tons / acre to pH desired		
pH	6.8	6.4	6.0
6.7	2.0	1.5	2.0
6.6	2.5	2.5	2.0
6.5	3.5	3.0	2.5
6.4	4.5	4.0	3.0
6.3	5.0	4.5	4.0
6.2	6.0	5.0	4.5
6.1	6.5	6.0	5.0
6.0	7.5	6.5	5.5
5.9	8.5	7.6	6.0
5.8	9.0	8.0	6.5
5.7	10.0	8.5	7.5
5.6	10.5	9.5	8.0
5.5	11.5	10.0	8.5
5.4	12.5	10.5	9.0
5.3	13.0	11.5	9.5
5.2	14.0	12.0	10.0
5.1	14.5	12.5	11.0
4.0	15.5	13.5	11.5
4.9	16.5	14.0	12.0
4.8	17.0	14.5	12.5

* Multiply these values by 2.24 to get t/ha.Source: Shoemaker, et. al. (1961)

Measure the pH of the soil buffer suspension on a pH meter after standardizing with known pH buffer solutions of pH 4.0, 7.0 and 9.2. Against the measured pH, find out the amount of lime required to bring the soil pH to a desired level. Make necessary correction to get the value of agricultural lime based on purity percentage.

Determination of phosphate fixing capacity of soil

Principle: Phosphorus use efficiency is generally very low in all soils due to the conversion of a part of the applied phosphates to unavailable forms through fixation. The extent of P fixation varies from soil to soil depending on the amount and type of clay minerals present. This leads to the necessity of making appropriate corrections in the recommended doses of phosphatic fertilizers. For determination of the P fixing capacity of soil, the method given by Waugh and Fitts (1966) is widely accepted and described below. In the procedure the soil sample is treated with graded doses of P and incubated for a period sufficiently to allow maximum fixation. The soil is then extracted and P determined. The recovery of P is subtracted from the added P and the retained or fixed P is calculated as per cent of the added quantity.

Instrumental: Colorimeter / spectrophotometer 2. Mechanical shaker

Reagents:

- **Olsen' reagents:** As in the case of determination of available P.
- **Ammonium molybdate solution:** As in the case of determination of available P.
- **Stannous chloride solution:** As in the case of determination of available P.
- **Activated charcoal (Darco G-60):** As in the case of determination of available P.
- **Standard P solution (2 mg P /L) :** As in the case of determination of available P.
- **Standard stock P solution (2500 mg P/ L):** Dissolve 4.720 g of fresh AR grade anhydrous monocalcium phosphate in about 300 ml of distilled water and dilute to 500 ml.
- **Dilute standard P solution:** Dilute 0, 1, 2, 3, 4, 6, 8, 10 and 15 ml portions of the stock P solution to 50 ml in volumetric flask to get the solutions containing 0, 50, 100, 150, 200, 300, 500, and 750 mg P/L.

Procedure: Weigh 2 g of air dry soil samples in duplicate in a series of 18 conical flasks of 100 ml capacity. Add 1 ml of each of the 9 dilute standard P solutions to the soil taken in different pairs of the flasks. These will results into conc. of 0, 25, 50, 75, 100, 150, 200, 250 and 375 µg of added P per g of soil. Plug the conical flasks with cotton wool and allow incubating at room temperature for 96 hrs. Extract the soil by Olsen' reagent for alkaline soil or Bray's reagent for acidic soil as in the case of determination of available P, take 5 ml of the clear and colorless filtrate and determine the P content.

Calculation: Available P (Extracted) content of soil without P treatment = A µg /g

Available P content of treated soil = B µg /g

Amount of P extracted out of added P = (B-A) µg /g

Amount of added P = C µg /g

Amount of P fixed by soil = C- (B-A) µg /g

$$\text{P fixation (\% of added P)} = \frac{C-(B-A)}{C} \times 100$$

Determination of potassium fixing capacity of soil

Principle: When soluble form of K is added to soil apart of it is readily converted or less or difficulty available form. This conversion is of great significance as it is not a net loss of K but is a temporary storehouse of K which becomes available later. Also this process prevents the loss of K by leaching. This fixation of K largely depends on the amount and type of dominant clay mineral, wetting and drying conditions of soil presence of other cations, available and added K contents, pH soil texture and moisture content of soil alternate wetting and drying cycles have been reported to increase the fixation of K, because of expansion and contraction of the mineral colloids, entrapping K^+ ions.

Instrumental: Flame photometer

Reagents

- **Neutral 1N ammonium acetate:** As described earlier.
- **Standard stock solution (1000 mg/L):** As described earlier.
- **Standard solution 'A' (10 mg K /L):** Dilute 1 ml of the stock solution to 10 ml.
- **Standard solution 'B' (100 mg K /L):** Dilute 10 ml of the stock solution to 100 ml.
- **Series of standard K solutions:** Prepare 0, 5, 10, 15,20,25,30, and 40 mg K /L solutions (100 ml each) for standard curve.

Procedure: Weigh 5 g of air dry soil samples in 9 to 11 pairs of 100 ml in conical flask or porcelain dishes. Add 0, 2.5 & 5.0 ml solution A and 1.0, 1.5, 2.0, 2.5, 3.0 and 5 ml of solution B each in duplicate. Add measure quantity of water so that each flask contains 10 ml of water respectively so as to maintain the quantity of liquid at 10 ml in each of the flasks. Stopper the flask with cotton wool and allow standing for 72 hrs at room temperature. Add 25 ml of neutral 1*N* ammonium acetate solution and shake for 5 minutes. Filter through Whatman No. 1 filter paper and determined K content in the clear filtrate, after drawing a standard curve as in the case of available K determination using a flame photometer.

Calculation:

Amount of ammonium acetate extractable K
in untreated soil = A

Amount of ammonium acetate extractable K
in untreated soil = B

Amount of recovered K
(Out of the added K) = B-A

Amount of K added = C

Amount of fixed K = C– (B–A)

$$\text{K fixation (\%)} = \frac{C-(B-A)}{C} \times 100$$

Comments:

1. In case the inflexion point on the curve is not sharp enough or is absent even if the procedure is repeated.
2. The mean of the amounts fixed from at least 3 to 4 graded doses may be worked out and used for calculating the percent fixation.

Plant analysis for P, K, secondary and micronutrients

Principle: For the release of mineral elements from plants tissue dry ashing and wet oxidation are the two widely adopted methods. Dry ashing is carried out usually at an ignition temperature of 550 to 600° C followed by its extraction in dilute HCl or H_2SO_4 for determination various elements. Ashing at temperature exceeding 600°C leads to considerable volatilization loss of P and K. To prevent this loss of P, dry ashing is carried out in presence of an alkali like Mg (NO_3) or M g (O Ac) $_2$ in alcohol. Alternately ignite in muffle furnace at 400-450 °C temperature is also available which prevents the loss of K. Volatilization loss of S and Cl also take places during ignition which can be prevented by adding Na_2CO_3. A part of P and micronutrients also get occluded causing a lot of error. For these reasons and for being comparatively more time taking dry ashing is only occasionally adopted. Wet oxidation employs oxidizing acids like HNO_3-$HClO_4$ tri-acids mixture or HNO_3-$HClO_4$ di-acid. Use of $HClO_4$ avoids the volatilization loss of K and provides a clear solution while H2SO_4 helps completion. $HClO_4$ on heating produces anhydrous $HClO_4$ which dissociates into nascent chlorine and oxygen, which in turn increase the oxidation efficiency at high temperature. Direct contact of $HClO_4$ with plant samples might leads to explosion and fire and hence the predigestion of samples in HNO_3 is preferred. Digestion with HNO_3-$HClO_4$ instead of the tri-acids mixture can also be adopted especially when S is also to be determined in the same digest. It is possible to directly extract originally bound and exchangeable K and Na from the plants tissues without oxidation by using 2N NH_4OAC-0.2 *N* Mg $(O\ Ac)_2$ for extraction and filtration. The wet oxidation method being easier, less time–consuming and convenient is given below.

Wet oxidation methods

Instrumental: 1. Water bath 2. Hot plate

Reagents:

- **Tri-acid mixture:** Mix AR grade conc. HNO_3, H_2SO_4 and $HClO_4$ in 10:1:4 ratio and cool.

Procedure: Transfer 0.5 or 1.0 g of dried and processed plant sample to a 250 ml conical flasks. Add 5 ml of conc. HNO_3. Keep a glass funnel on the flask, place it on a water bath and heat at 100°C for about 30 minutes. Shift the flasks to a hot plate and heat at 180 – 200°C. Measure temperature in a flasks containing glycerol kept on the hot plate. Continue boiling until near to dryness, but not drying completely. Cool and add 5 ml of the mixture is largely volatilized. If the contents are still brown cool a little and add 3-4 ml of the tri-acid mixture and continue the digestion as described above. This is rare as there is still sufficient $HClO_4$ to oxidize the charred materials. Remove the flasks when only moist, clear and white contents are left. The entire quantity of $HClO_4$ has volatilized by this stage. Cool and adds about 50 ml of distilled water or double DDW if micronutrients ate to be analyzed. Filter into 100 ml flasks giving washings to make the volume to 100 ml. Use filtrate for analysis.

Di-acid digestion

Instrument: Hot plate

Reagents:

- Conc. HNO_3 (AR grade)
- 60 % $HClO_4$ (AR grade
- Approx. 2 N HCl (AR grade

Procedure: Weigh 0.5 or 1.0 g of dried and processed plant samples in a 100 ml or 150 ml conical flask. Add 10 ml of conc. HNO_3 place a funnel on the flask and keep for about 6-8 hrs or overnight at a covered place / chamber for predigestion. After predigestion when sample is no more visible add 10 ml of conc. HNO_3 and 2-3 ml of $HClO_4$. Keep on a hot plate in acid proof digestion chamber having fume exhaust system and heat at about 100 °C for first one hrs and then raise the temperature to about 200°C. Continue digestion until the content become colorless and only white dense fumes appear. Reduce the acid contents to about 2-3 ml by continuing heating at the same temperature. Do not allow drying up. Remove from hot plate, cool and add 10 ml of dilute and perfectly colourless HCl. Warm slightly and filter through Whatman No. 42 filter paper in to a 100 ml volumetric flask. Give 3-4 washings of 15-20 ml portions of distilled water and make the volume to 100 ml.

Total Nitrogen in plant: Total nitrogen in plant is determined commonly by the Kjeldhal method and by colorimetric procedure. This involves

- Digestion of the sample to convert the N compounds in the sample to the NH_4^+ form and
- Determination of NH_4^+ in the digest

Colorimetric procedure (Novozamsky et. al., 1974): A colorimetric can also be followed either manually or by using auto analyzer system for determination of NH^4-N in the plant digest.

Reagents

A. For digestion

- **Acid mixture (sulphuric acid solution 0.7 *M*):** add 40 ml H_2SO_4 (sp.gr. 1.84) to 500 ml distilled water in a volumetric flask of 1 litre. Cool and make up to the marks.
- **Hydrogen per oxide 30%**

B. For colour development

- **Sodium hydroxide 1 M:** Dissolve 40 g of NaOH in one litre of distilled water.
- **Alkaline phenolate solution:** Dissolve 22 g of phenol in 250 ml of 1M NaOH and make up the volume to 1 litre. The solution is stable for at least one week. Sodium nitropruside in 1 litre of distilled water.
- **Disodium EDTA solution 4% w/v:** Dissolve 4 g disodium EDTA in a few ml distilled water in a 100 ml volume flask and make up the volume up to the marks.
- **Phosphate buffer solution:** 13.35 g of $Na_2HPO_4.2H_2O$ in distilled water. Add 50 ml of 1M NaOH and make up the volume to one litre. If necessary, adjust the pH to 12.
- **Sodium hydrochloride solution:** Dilute 20 ml of sodium hydrochloride solution (approx. 1 M hydrochloride solution in 0.1 M NaOH to 100 ml. The final solution should contain about 0.7 % of active chlorine.
- **Mixed reagent I:** Mix 100 ml of alkaline phenolate solution with 200 ml of sodium nitropruside solution and 10 ml of EDTA solution. Prepare immediately before use.
- **Mixed reagent II:** Mix 400 ml of phosphate buffer with 100 ml sodium hydrochloride solution.

Procedure

A. **Digestion:** Weigh accurately about 0.3 g of air dry, finally ground plant material in a 50 ml volumetric flask. Add 3.3 ml of the acid moisture; mix well waits at least one hrs for the reaction to proceed use 3.3 ml of the acid mixture as the blank.

Heat moderately on a hot plate and swirl gently now and then to minimize foaming. should foam enter the neck of the flask. Add 1 or 2 drops of hydrogen per oxide and raise the temperature to about 280°C continue to add 5 drops of hydrogen peroxide at 10 minutes interval until the resulting solution remains clear for 10 minutes at 280 °C.

B. **Colour development**: Take 0.5 – 1.0 ml of digest in 25 ml volumetric flasks and add 3 ml of mixed reagent I and 5 ml of mixed reagent II. Mix after each addition and make up to the volume with distilled water. Measure the colour intensity at 630 nm after 90 minutes.

C. **Preparation of standard curve:** Dissolve 0.3820 g of pure ammonium chloride in 1 litre distilled water. This solution contains 100 mg of ammonium nitrogen in one litre. Ten ml of this is diluted to 100 ml to get 10 ppm solution. 0.0,0.5,1.0,1.5,2.0,4.0,ml of this 10 ppm solution is pipette out in to 25 ml volumetric flasks and the colour is developed and read as above. The absorbance at 630 nm is plotted against conc. of ammonium nitrogen.

Comment

1. Since nitropruside is unstable in an alkaline medium. It is considered necessary to prepare a fresh solution before the analysis. A relatively high concentration of this reagent has been chosen to increase the reaction rate of the colour reaction.
2. Once developed the blue colour is a stable for at least 24 hrs. The colour is virtually completely developed in 60 minutes and the method has a molar extinction coefficient of 15,000. Beer's law is following the range of the standard series.
3. The interference of some cations was examined at concentrations which were a multiple of those actually present in plant materials.

Phosphorus: Phosphorus in an aliquot can be determined by following the methods based on molybdophosphoric blue colour developed by reduction of the heteropoly complex, or vandomolybdophosphoric yellow colour method. In the heteropoly complexes, P acts as the central coordinating atom to form 12- form heteropoly acids with molybdate and the oxygen of molybdate is substitute for that of phosphate as shown below;

$$H_3PO_4 + 12\ H_2MoO_4 \longrightarrow H_3\ P\ (Mo_3O_{10})_4 + 12\ H_2O$$

Selection of appropriate method for P determination largely depends on the P content and the interfering ions present in the sample as the sensitivity varies from method. The vandomolybdophosphoric yellow colour method given below is suitable for P determination in plant extracts. In this method, through the sensitivity (1 to2 mg/L) is about one-tenth of those of the blue colour methods, they yellow colour developed is more stable no reluctant is needed and no interference of other ions take place even if present up to 1000 mg/L concentration. HCl, H_2SO_4 or $HClO_4$ can also be used; through HNO_3 system is often preferred.

Instrument: Spectrophotometer or Colorimeter.

Reagents

- **Vanadate-molybdate reagent**: Prepare solution A by dissolving 25 g of ammonium molybdate in about 400 ml of warm water. Prepare solution B separately by dissolving 1.25 g of ammonium metavanadate in about 300 ml of boiling water cool it and add 250 ml of concentration HNO_3. Cool again at room temperature. Now add solution A to solution B and dilute to one litre.

- **Standard P solution**: prepare solution containing 100 mg P / L by dissolving 0.439 g of dried $K_2H_2PO_4$ in water, acidifying with 25 ml of 7 N H_2SO_4 and making the volume to 1 L.

Procedure: Transfer a suitable volume not more 30 ml to 50 ml volumetric flak so that it contains 0.05 to 0.1 mg of P and the acid equivalent is between 0.6 and 1.6 *N* in the final volume of 50 mL. Add 10 ml of the Vanadate-molybdate solution and dilute to 50 mL with water. Mix well and read the colour intensity after 10 minutes using blue filter (420 nm).Run a blank without P solution simultaneously. Take 0.50, 0.75, 1.0, 1.25, 1.50, 2.00 and 250.0 mL of the 100 mg P /L solution in 50 ml volumetric flasks and develop colour in identical manner. Prepare standard curve by plotting P concentrations on X-axis and per cent transmission/ colorimeter readings on Y- axis.

Calculate P content of sample using the standard curve.

Calculation

$$\text{P content of sample (\%)} = \frac{A}{100V}$$

Where A stands for P concentration in µg as read against the simple reading on the graph and V stands for quantity of test sample taken (mL) for colour development out of the 100 mL acid digest made from 1 g sample.

Potassium: Potassium in the acid digest of plants samples can be determined using flame photometer exactly in the same manner as described for available K in soil. Depending on the concentration of K in the plant samples, the digest can be used either directly or after dilution for flame photometric determination. For example wheat grain digest can be taken up directly while straw samples are diluted 10 times for this purpose. Alternatively, use AAS, which provides better precision.

17- Calcium and Magnesium: These elements can be determined in the di-acid digest of plant sample either using AAS or versene titration. Calcium can be estimated on a flame photometer also, through the sensitivity is less than that for K and Na.

18- Sulphur: Since, dry ashing leads to volatilization loss of S present in the organic combination and the wet oxidation based on tri acid mixture includes H_2SO_4 both of these methods can not be used for S determination in plant samples. Therefore, HNO_3-$HClO_4$ digest of plant is conveniently used. Turbidmetrically method as described for soil sulphur or the barium chromate colorimetric method (Palasker *et. al.* 1981) can be used.

19-Iron, Zinc, manganese and copper: The above mentioned four metallic cations can be determined in the di-acid digest of plant tissues using AAS. Dithizone method (Shaw and Dean, 1952) for Zn and Cu ortho-phenanthroline method for Fe and Na-paraperiodate method for Mn can be used as an alternative colorimetric method.

20-Boron and Molybdenum: For B determination in plants tissue, follow the steps given below: Proceed for dry ashing as described earlier with 0.5 g samples. Extract the dry ash with 10 ml of 0.1N HCl. Filtrate or centrifuge the suspension to a clear state. Pipette 1 ml aliquot from the clear solution and proceed for B determination by Azomethine-H method as described earlier.

For the determination of Mo in plant tissues, ashing and fusion of the sample is done as given below:

Take a suitable sample (1 to 10 g) in a platinum dish and ash in an electric furnace at 500 °C. Fuse the ash with 2 g of anhydrous Na_2CO_2 at 1000 oC in the electric furnace ensuring complete contact of the entire ash with the flux. Cool the dish and drop the cake into a 250 ml beaker by inverting the dish. If the cake does not dispatch place the dish in beaker and disintegrate the cake in 100 ml water containing 2% ethanol by volume. Measure the suspension in a cylinder and filter through Whatman No.42 filter paper in a 250 ml breaker. Measure the undiluted clear filtrate and use for Mo determination as described earlier. Calculate on the basis of the former volume.

Critical level of nutrients: This is well established concept for diagnosing nutritional problems in plants.

Table: General sufficiency or optimum range of nutrients elements in plants.

Nutrients	Sufficiency or optimum range
Macro-nutrients (%)	
N	2.0-2.5
P	0.2-0.5
K	1.0-5.0
Ca	0.1-1.0
Mg	0.1-0.4
S	0.1-0.3
Micro-nutrients (µg / g)	
Zn	20-100
Fe	50-250
Mn	20-300
Cu	5-20
B	10-100
Mo	0.1-0.5
Cl	2000-20,000

Source: Soil Plant and Water analysis-Dhyan Singh *et. al.*, 2005

It indicates the single point on the curve relating nutrients content to crop yield. This level or range may be considered as the "just sufficient" or "minimum" for the highest yield. Obviously, a level below it will lead to decreased yield. An adequate range also follows above the critical level, which is considered as the goal to be achieved while making recommendations. However, there could be even the upper critical level of nutrient in the end of adequate range in the case of some nutrients, which may adversely affect the yield. Nutrient balance in plant can be judged from the ratio of various nutrients. Commonly used ratios which reflect some nutritional antagonism are N/S, K/Ca, K/Mg, Ca^+ Mg/K and N/P.

Analysis of Fertilizers and Manures: Analysis of fertilizers and manures is required to know their composition and quality. The analysis of fertilizers is either physical or chemical. Various methods have been prescribed for analysis of fertilizers and manures of different parameters in many standards books, some of which are applicable in different country.

Sample collection and preparation: While sampling procedure is required for samples that are representative of the whole lots is of utmost important in ensuring the accuracy of analytical results. The prerequisite of good sampling is to draw best possible representative samples with minimum contamination and in a best possible natural condition. This require first subdivision of huge big lots into smaller lots than selection of specific representative begs without any elements of bias and finally drawing samples from the selected begs to make representative samples. The samples –

- Should not be taken at a place which exposed to the rain or sun.
- The sampling instrument should be clean and dry and the material should be free from contamination.
- Sampling should be mixed as thoroughly as possible by means to make it homogenous.
- The sampling should be kept in suitable, clean, dry and air tightly glass in a thick gauge polythene bag.

To prepare the final test samples from bulk, break all lumps and pulverize the material so that it passes through a sieve (1 mm size). Store the samples in an air tightly bottle and label it with the details of sampling. Sampling from bags, withdraw equal quantities of materials from the top, middle and sides of the each bag and mix to form a composite samples.

For begged fertilizers, use a slotted single or double tube probe with solid cone tip made of stainless steel or brass.

For bulk fertilizers in truck use a stainless steel or brass sampling cup with inside trucks use a stainless steel or brass slotted double tube probe with solid cone tip with approx.

For quality control purpose depending upon the lot size 1 to 10 begs are selected from a lot of 100 tonnes details. The identified sampling bags should be laid horizontally and mixed thoroughly. The sampling probe should be inserted diagonally from one corner to another corner keeping the slit down and rotate while withdrawing. The fertilizer is emptied in a container or clear hard surface. Two piercing are required from each beg. All samples thus drawn are to be mixed together and reduced by the process of quartering or riffling to about 1.2 kg. Out of this, three reference samples are to be made and kept in clearing right container with identification marks. The sample received in the laboratory is ground quickly to avoid any absorption of moisture and all materials made >1mm by repeated less than sieving and grinding. However, for the fertilizers whose particle size is already less than 1 mm, only thoroughly mixing needed to make a homogenous mass.

Determination of Moisture: The fertilizers posse three forms of moisture 1) water of crystallization 2) water of absorption / adsorption 3) free water.

Water of absorption/adsorption and free water may also present in manures. Based on the nature of manures and fertilizers, the different methods for moisture determinations are oven dry method, vacuum desiccator methods, Karl Fisher titration method and the evaporation method. The all methods have been described in below.

Oven Dry Methods: This method is used for stable substances which do not give volatile substance at 100°C like manure, mixed fertilizers, fertilizers other than urea and ammonical fertilizers. Urea and DAP CAN and nitrophosphate etc. are not analyzed by this method.

Procedure: Weigh 2 gram of sample in a clean pre-weighed porcelain or silica or glass dish and heat in an oven for about 4 -5 hour at 100 °C ± 1°C to a constant weight. Cool it and weigh accurately. For sodium nitrate and potassium salts heat up to 130 ± 2 °C. Calculate the moisture as percentage loss in weight at particulate temp used.

Calculation

$$\text{Moisture } \% = \frac{100\,(B - C)}{(B - A)}$$

Where, A= Weight of empty bottle, B= weight of bottle + sample (before drying), C= Weight of bottle + sample (after drying)

The moisture is reported on fresh weight basis only.

Vacuum desiccator Method: This method determine free water content of fertilizers (suitable for Ammonium Chloride, Ammonium nitrate DAP urea, CAN potassium nitrate etc.)

Procedure: Weigh accurate 5.0 samples in a clean pre-weighed porcelain or silica or glass dish. Place the dish in vacuum desiccator over previously boiled concentration H_2SO_4 for 24 hrs. Now weigh again.

Calculation:

$$\text{Moisture } \% = \frac{100\,(B - C)}{B - A}$$

Where, A= Weight of empty bottle, B= weight of bottle + sample (before drying), C= Weight of bottle + sample(after drying)

Karl Fisher Titration Method: The Karl Fisher method determines total water including water of hydration. This method is based on the reaction of water with iodine SO_2 and pyridine(C_5H_5N) solution. The pyridine prevents the loss of SO_2 from the reagents. It is suitable for urea, CAN, nitrophosphate etc.

Reagent: Methanol, Sodium tartarate and Karl Fischer Reagent, KF (Pyridine free)

Procedure: Standardization of KF reagent: Add 25 ml methanol to the titration vessel to dip the electrodes and titrate with KF reagent to a pre-set end point. Accurate weigh 0.2 g of sodium tartarate dihydrate transfer to titration vessels stir and titrate with KF reagent to the same end point (V_1)

Titration of the sample: Add 25 ml methanol to the titration vessel and titrate with KF reagent. Transfer an accurate weighed sample containing approx. 30 mg of water to the titration vessel. Stir and titrate with KF reagent to the same end point (V_2).

Calculation:

$$\text{Mg } H_2O \text{ / ml of KF} = \frac{156.6 \times \text{wt. sodium tartarate}}{\text{ml KF reagent used (V1)}}$$

$$\text{Moisture \%} = \frac{V \times F}{W \times 10}$$

Where, V= ml of KF reagent used (V_2), F= mg water / ml KF reagent, W = wt. of sample (g)

Comments and suggestions:

- All the apparatus should be free from moisture and use of silica gel is normally preferred.
- KF reagent (A+B) is also being manufactured and marketed separately.
- Now the pyridine free base has also been developed containing single solution of Iodine and SO_2 making analysis easier.

Estimation of Nitrogen: The forms of N normally required to be determined in different fertilizers are (i) total N (ii) Ammonical–N (iii) Ammonical and Nitrate–N (iv) Urea (Amide) N and (v) Water Insoluble-N. Determination of N in fertilizers and manures involved three following major step:

- **Digestion:** By this method, the N present in sample is converted to the ammonical form.
- **Distillation:** By this method, there are two purposes like treatment with strong alkali to liberate ammonia and absorption of NH_3 in known excess quality of standard acid.
- **Titration**: this involves volumetric determination of ammonia through back titration of excess acid.

The nitrogen in ammonical nitrogenous fertilizer is determined by following method. Ammonical N determination by distillation with an alkali (NaOH or MgO) has been described below:

Principle: The NH_4-N is liberated from ammonical fertilizers by alkali (45% NaOH) which is then absorbed in known quantity of standard acid (4% boric acid). The amount of such NH_3 liberated is calculated from the volume of standard acid (0.1N HCl or 0.1N H_2SO_4) consumed with mixed indicator (Methyl Red indicator).

Reagents:

- Std. 0.1 N NaOH,
- Std. 0.1 N HCl or H_2SO_4
- Zinc granules
- NaOH 45% solution
- MgO freshly ignite carbonate–free MgO
- Methyl red indicator solutions

Procedure: Weigh accurately 1.0 g of the given sample and dissolve in water. Filter, if required and make up volume to 250 ml, and transfer 25 ml aliquot then add 300 ml water and 5 ml of 45% NaOH or 2 g MgO to the Kjeldhal flask in such a way that it runs down from the neck to the bottom without mixing and fit the cork immediately. Mix the solutions thoroughly by swirling and light the gas burner and start heating at once to avoid any danger of sucking back and immediately connect to distillation unit and continue distillation till about 150 ml of the distillate is collected then remove the conical flask and rinse the outlet tube with distilled water. Titrate the content with standard alkali using methyl red as indicator till blue colour disappears and tends to turn pink at the end point. Carry out the blank with 250 ml of water and 25 ml of 45% NaOH solution without the addition of ammonium sulphate.

Calculation

Weight of sample	=	1g
Volume of fertilizers sample	=	100 ml
Aliquot taken from fertilizer sample	=	10 ml
Normality of Std. HCl used for neutralization of ammonia	=	0.1 N
Titration volume (T)	=	Sample value – Blank value
Since, 1000 ml of 1 N HCl	=	14 NH_3 = 14 g N
1 ml of 0.1 N HCl = 0.0017g NH3	=	0.0014 g N
Weight of N in aliquot	=	0.0014 x T

$$\text{Weight of N in 100 ml fertilizer solution} = \frac{0.0014 \times T}{10} \times 100$$

% N in fertilizer $= \frac{0.0014 \times T \times 100}{100} \times 100$

Determination of ammonical and Nitrate N: The methods recommendation for determination in fertilizers containing only NH_4-N and NO_3-N, the ammonical nitrogen can be determined by the direct distillation of ammonia with 45 % NaOH. Whereas the nitrate can be reduced by zinc and ferrous sulphate in alkali medium or zinc and Al in alkaline medium on the chemicals used to reduced nitrate-N into ammonia two methods are prescribed below:

- Ferrous sulphate zinc soda method
- Devarda alloy method

Both these methods can be used where whole of N is present in inorganic form (NH_4 and NO_3). It is not applicable if the samples contain urea, or OM or Ca cyanamide.

The Devarda alloy method is described below.

Principal: Devarda alloy contains 50% Cu, 45% Zn, 55 Al which reduces NO_3 to NH_3 in a alkaline medium whereas Cu increase bitterness of the alloy and Zn itself prevents bumping during distillation.

$Zn + 2NaOH \longrightarrow Na_2ZnO_2$ (Sodium zincate) $+2\ H^+$

$Al + 2NaOH \longrightarrow Na_2AlO_2$ (Sodium aluminate) $+2\ H^+$

$NO_3 + 9\ H^+ \longrightarrow NH_3 + 3\ H_2O$

Reagent

- Std. 0.1 N NaOH
- Std. 0.1 N HCl or $H_2\ SO_4$
- NaOH-45% solution
- Methyl red indicator solution
- Devarda alloy (50% -Cu, 45% Zn and 5% Al.)

Procedure: Weigh accurately 1.0 g of the sample and dissolve in distilled water, filter if required and make up the volume to 250 ml then transfer 25 ml aliquot to a distillation flask and add 300 ml water, 3 g of finally powered Devarda alloy and 5 ml of 45% NaOH solution carefully along the side of flask so that it does not mix at a once with the contents. Immediately connect to distillation unit immersing the out let tube of condenser in 20-25 ml std acid containing 4% boric acid solution with 2-3 drops methyl red. After 15-20 minute mix the contents of distilling flask by rotation. Now heat slowly first and then at a rate to yield 250 ml distillation in an hours. Collect distillate and titrate with std acid (0.1 N HCl or H_2SO_4). Carry out the blank.

Calculation:

$$NH_4\text{-}N + NO_3\text{-}N\% \text{ in sample} = \frac{0.001 \times T \times 100 \times 100}{10}$$

Where, T = sample value – Blank value

If N is to be expressed as NO_3^+ the factor 0.0062 is to be used instead of 0.0014.

Determination of Total Nitrogen: This method is suitable for the materials having organic N with nitrate in compost and manures (FYM).

Principal: The materials is digested with H_2SO_4 and salicylic acid in the presence of K_2SO_4, $FeSO_4$, $CuSO_4$ and $Na_2S_2O_3$ (sodium thiosulphate) or zinc dust. The K_2SO_4 raises the boiling point of H_2SO_4 and $CuSO_4$ is act as catalyst. The salicylic acid is reduced to amino compound by $Na_2S_2O_3$ (sodium thiosulphate). The ammonia is, then, estimated by the method of Kjeldhal method as described previously.

Reagents

- Sulphuric acid-Salicylic acid mixture: add 1 g of salicylic acid to 30 ml of sulphuric acid.
- Sodium thiosulphate or zinc dust
- Conc. H_2SO_4
- Std 0.1N HCl solution
- NaOH solution 45%
- Mixed indicator: dissolve 0.5 g bromocresol green and 0.1 g methyl red in 100 ml of 95 % ethanol.
- Boric acid 4%: dissolve 40g H_3BO_3 in 1 litre DSW by heating gently and add 5 ml of mixed indicator in it and adjust pH of boric acid to 4.5 by using NaOH or HCl (colour turns from blue to slightly pink).

Procedure: For digestion process, weigh 1 g sample or prepare 0.1 % solution for urea and take 10 ml of this in Kjeldhal flask. Add 10 g of salt mixture and 30 ml conc. H_2SO_4. Shake the contents and heat gently until frothing ceases. If required, add small amount of paraffin to reduce frothing. Raise the heat until solution become whitish upon complete digestion. Cool and add slowly with intermittent shaking about 200 ml of water and cool again. In distillation process, immediately connect to distillation unit immersing the out let tube of condenser in 50-100 ml conical flask containing 4% boric acid solution with 2-3 drops mixed indicator and mix by swirling. At least 150 ml of distillate has been collected if any NH_3 is still coming out test with red litmus paper. Titrate with std 0.1 N HCl till blue color disappear. Carry out the blank.

Calculation

$$\text{N\% in sample} = \frac{0.001 \times T \times 100 \times 100}{\text{wt of sample}}$$

Where, T = sample value – Blank value

If N is to be expressed as NO_3^+ the factor 0.0062 is to be used instead of 0.0014

Precautions

- In case of flame or light goes out off during distillation immediately remove the receiver to prevent the sucked back in to the Kjeldhal flask.
- If there is severe bumping, regulate the flame or heater.
- At the end without putting off the burner, disconnect the tube from the condenser.
- This process is suitable if nitrates are present in the sample because they would be lost as HNO_3 during digestion.

Determination of Phosphorus: The different forms of phosphates required to be analyzed in different fertilizers are-

- Total P_2O_5
- Water soluble P_2O_5
- Available P_2O_5 or neutral ammonium citrate soluble P_2O_5
- Ammonium citrate insoluble P_2O_5
- Citric acid soluble P_2O_5

Determination of Total phosphates as P_2O_5: The phosphorus in fertilizers and manures can be determined by following methods-

- Volumetric ammonium Phospho molybdate
- Gravimetric ammonium Phospho molybdate
- Volumetric quinolinium Phospho molybdate
- Gravimetric quinolinium Phospho molybdate
- Spectrophotometric vanado Phospho molybdate

The Volumetric quinolinium Phospho molybdate method is suitable for all types of fertilizers, manures and fertilizer mixture because it is very accurate, simple and free form interferences hence, normally adopted by analytical laboratories.

Principal: Phosphorus is converted to orthophosphate as quinolinium phospho molybdate ($C_9H_7N\ 3H_3PO_4.12MoO_3$) in boiling medium by Quimociac reagents. It can either be weighed gravimetrically or titrate with alkali. It is applicable to solutions containing both orthophosphate and polyphosphate. The orthophosphate also form yellow colour molybdo-phosphoric acid by reacting with metavanadate which can be used for Spectrophotometric analysis. Ca, Fe Mg alkali metals and citrates do not affect analysis in gravimetric methods. The citrate in the reagent complexes the ammonium ions thus, preventing interferences form precipitation of ammonium phospho molybdate by ammonium salts usually present in mixed fertilizers. The citrates also lessen interferences from soluble silica.

Reagents

- Conc. HNO_3
- Conc. HCl and Conc. H_2SO_4
- Sodium nitrate / potassium nitrate-AR Grade
- Perchloric acid -70-72 %
- **Magnesium nitrate soluble (9%)**: dissolve 90 g of phosphorus free mg $(NO_3)_2$ in water and dilute to 1 litre.
- **Quimociac reagent**: Dissolve 60 g citrate acid in mixture of 85 ml HNO_3 and 150 ml water and the cool. Take 70 g of sodium molybdate dihydrate and dissolve in 150 ml water. Gradually add sodium molybdate solution to citrate acid-nitrate mixture with stirring. Dissolve 5 ml synthetic quinoline in mixture of 35 ml HNO_3 and 100 ml of water. Gradually add this solution to molybdate citric nitric acid solution, mix and keep it stand for 24 hours and filter. Add 280 ml acetone, dilute to one Litre with water and mix well. Store in polyethylene bottle.

Sample preparation procedure: According to the nature of fertilizers the sample should be treated with one of the following methods.

- **For materials containing high organic matter**: Take 1 g sample in evaporation dish and add 5 ml of $Mg(NO_3)_2$ solution and evaporate to dryness and boil gently until the organic matter is destroyed and red brown fumes cease to appear.
- **For materials containing small quantity of organic matter**: Take 1 g sample in 50 ml beaker and add 30 ml of HNO_3 and 5 ml HCl and evaporate to dryness and boil gently until the organic matter is destroyed and red brown fumes cease to appear.
- **For materials and fertilizers mixtures containing much Fe and Al phosphate** : Take 1 g sample 30 ml HCl and 10 ml HNO_3 and boil gently until the organic matter is destroyed and red brown fumes cease to appear.

Cool the solution prepared by any of the three above methods, dilute to 250 ml volume, mix and filter through dry filter if required.

Water soluble phosphates

Principle: The water soluble portion of P_2O_5 in the sample is filtered and phosphate is obtained in the filterrate. The soluble orthophosphate is precipitated by quimociac reagent as quinolinium phospho molybdate (C_9H_7N $3H_3PO_4.12MOO_3$) which is dried at 250°C and weighed. This method is applicable to all fertilizers included fertilizers mixtures.

Reagents:

- **Quimociac reagent**: Dissolve 60 g citrate acid in mixture of 85 ml HNO_3 and 150 ml water and the cool. Take 70 g of sodium molybdate dihydrate and dissolve in 150 water. Gradually add sodium molybdate solution to citrate acid-nitrate mixture with stirring. Dissolve 5 ml synthetic quinoline in mixture of 35 ml HNO_3 and 100 ml of water. Gradually add this solution to molybdate citric nitric acid solution, mix and keep it stand for 24 hours and filter. Add 280 ml acetone, dilute to one Litre with water and mix well. Store in polyethylene bottle.
- **Nitric acid** –AR grade: (Phosphorus free)

Preparation of sample: Weigh 1.0 g prepared sample on filter paper No.1 or 5 in a funnel in a 250 ml volumetric flask. Using a fine stream of water directed in a circular path around the entire periphery of the filter paper, wash the sample with 10-15 ml portions until 240-250 ml filtrate have been collected within one hours. Ensure that the water and sample are thoroughly mixed with each washing and allow each portion of water to pass through the filter before adding the next portion if the filtrate if turbid add 1-3 ml of HNO_3 and dilute to 250 ml and mix.

Methods: Pipette out usually 25 ml an aliquot in to a 500 ml beaker and dilute to 50 ml if required and add 10 ml HNO_3 (1+1) and boil gently for 10 minutes to hydrolyze non-orthophosphate then cool and dilute to 100 ml with water. After this add 50 ml quimociac reagent shake well, cover with watch glass place on hot plate and boil for 1 minute and remember that not to use open flame. Cool to room temperature, swirl carefully 3-5 times during cooling. Allow the precipitate to settle and filter through sintered glass crucible No 4 and after complete transfer wash the precipitate 5 times using 25 ml of water for each wash, Use vacuum pump and suck dry between washings. Place the crucible in a drying over and dry at 250°C for 30 minutes and cool in a desicator and weigh as ($C_9H_7N_3$ H_3PO_4.$12MoO_3$). Carry out reagent blank and subtract the weight of reagent blank precipitate

Calculation

$$\text{Water Soluble Phosphates} = \frac{3.207 \times A}{W} \times df$$

Where, A = weight of precipitate (after subtraction of reagent blank precipitate)

W = Weight of sample in grams

df = Dilution factor in present case (250/50=10)

Determination of potassium: Potassium in all fertilizers is found mostly in water soluble form. Methods recommended for determination of potassium are

1. Volumetric STPB methods
2. Gravimetric STPB methods
3. Gravimetric Perchloric acid method
4. Gravimetric chloroplatinate methods
5. Gravimetric cobaltinitrite method
6. Volumetric cobaltinitrite method
7. Flame photometric method.

The Volumetric cobaltinitrite method and mainly STPB volumetric methods are most commonly used in the laboratory due to its simplicity and accuracy.

Sodium tetra phenyl Boron (STPB) methods

Principle: Potassium is extracted first from fertilizer sample with ammonium oxalate. The K in solution is precipitated with an excess of STPB as potassium tetra phenyl boron. The excess of STPB is back titrated with Benzalkonium chloride or zephiran chloride or quaternary ammonium salt using Clayton yellow as indicator.

$$Na\ [B\ (C6H5)_4 + K^+ \longrightarrow K\ [B\ (C6H5)_4] + Na^+$$

The interference of NH_4^+ takes place in precipitation hence it is first complexes with formaldehyde under slightly alkaline condition before precipitation of K. The formaldehyde does not complex quaternary ammonium compounds and the sulphate and chlorides does not interfere with the titration.

Preparation of sample solution

- **NPK complexes and NPK fertilizer mixtures:** Place 2.5 gm of prepared sample in 250 ml volumetric flask add 125 ml water add 50 ml of 4% ammonium oxalate solution add 1 ml diglycol separate to prevent foaming if needed. Boil for 30 minutes add light excess of NH_4OH and after cooling dilutes to 250 ml mix and pass through dry filter No. 12 or equivalent.
- **If the presence of organic matter:** Weigh accurately 5 gm of the sample in an evaporating dish and incinerate in muffle furnace to carbon free ash. Heat the ash with 10 ml of concentration HCl with 10 minutes. Cool and transfer into a beaker. Wash the dish with distilled water several times and collect the washings in the beaker. Dilute with water to about 150 ml and heat to boiling. Add powered barium hydroxide to make the contents just alkaline. Cool make volume up to 250 ml and evaporate to dryness in a porcelain dish. Heat the residue gently over a low flame with a few ml of conc. HNO_3 till all ammonium salts are expelled wet the residue with conc.

HCl and again evaporate to dryness add dilute HCl and digest on a water bath. Filter and wash free from chloride. Use the filtrate for precipitation of K

- **Straight potassium fertilizers (MOP,SOP, potassium Magnesium sulphate and Kainite):** Dissolve 2.0 gm sample and dilute to 250 ml with distilled water.

Reagents

- **NaOH solution (20%):** Dissolve 20 gm NaOH in 100 ml distilled water.
- **Formaldehyde solution (37%)**
- **STPB solution (approx. 12%):** Dissolve 12 gm sodium tetra phenyl boron in approx. 800 ml water. And add 20-25 gm Al $(OH)_3$, stir for 5 minutes and filter through Whatman No.42 paper or equivalent into a 1 litre volumetric flask. Rinse the beaker sparingly with water and add to filtrate. Collect entire filtrate and add 2ml 20% NaOH solution dilute the volume with water and mix. Let it stand for 48 hrs and standardize. Adjust so that 1 ml STPB = 1 % K_2O. Store at room temperature.
- **Benzalkonium chloride or zephiran chloride or quaternary ammonium chloride solution (appox. 0.625 %)**: dilute 50 ml of 1.28% Benzalkonium chloride (BCA) to 1 litre with water, mix and standardize. If other, concentration is used adjust volume accordingly.
- **Clayton yellow, 0.04%**: dissolve 40 mg in 100 ml water.
- **Ammonium oxalate solution** $(NH_{4)2}\ C_2O_4$ (4%).

Standardization of solutions

1. **Benzalkonium chloride**: Take 1.0 solution in STPB solution in 250 ml Erlenmeyer flask add 20-25 ml water, 1 ml 20%NaoH, 2.2 ml HCHO, 1.5 ml of 4% ammonium oxalate and 6-8 drops of Clayton solution using so that 10 ml semi micro burette. Adjust Benzalkonium chloride solution using 10 ml semi micro burette. Adjust Benzalkonium chloride solution so that 2.0 ml = 1.0 ml STPB solution.
2. **STPB solution**: Dissolve 2.5 gm of KH_2PO_4 in water in 250 ml volumetric flask add 50 ml 4% ammonium oxalate solution dilute to volume with water and mix. Transfer 15 ml aliquot (51.92mg K_2O or 43.10 mg K) in 100 ml volumetric flask and add 2 ml of 20% NaOH, 5 ml HCHO and 43 ml STPB reagent. Dilute to volume with water and mix thoroughly. Let it stand 5-10 minutes and pass through dry filter paper No. 42 transfer 50 ml aliquot of filtrate in the 250 ml Erlenmeyer flask, add 6-8 drops of indicators (Clayton yellow) and titrate excess

STPB with Benzalkonium to pink solution to pink end point. Factor can be calculated as below:

$$F= \frac{34.61}{(43.0\ -\ \text{ml Benzalkonium})} \times K_2O\ /\ \text{ml STPB reagent}$$

(This factor applies to all fertilizers if 2.5 gm sample is diluted to 250 ml and 15 ml aliquot is taken for analysis. If results are to be expressed as K rather than K_2O substitute 28.73 for 34.61 in calculating the value of F)

Procedure : Take 15 ml aliquot of sample solution as per prepared (1) and (2) above to 100 ml volumetric flask and add 2 ml 20 % NaOH and 5 ml HCHO and add 1 ml std. STPB solution for each 1% K_2O expected in sample plus additional 8 ml excess to ensure complete precipitation. Dilute to volume with water, mix thoroughly and let it stands for 5-10 minutes and pass it through dry filter paper Whatman No.12 or equivalent. After filter transfer 50 ml filtrate to 250 ml Erlenmeyer flask. And add 6-8 drops of indicator (Clayton yellow) and titrate excess STPB with standards Benzalkonium solution to pink end point.

Calculation

$\%\ K_2O$ = (ml STPB – Benzalkonium used) × F

Where, F = $\%\ K_2O$ / ml of STPB reagent

Comments and suggestion

- Do not premix the NaOH and formaldehyde solution as such mixtures are not stable and loose their power to complex NH_4^+.
- Acetone is a good solvent for the tetra phenyl boron precipitate; therefore, its use also helps in cleaning glassware.
- While calculating the factor as % K_2O / ml of STPB, the figure of 34.61 is the actual % age of K_2O present in std. KH_2PO_4.

Determination of Micro-nutrients in fertilizers

Spectrophotometric or atomic absorption spectrophotometer (AAS) are often for analysis of various elements in straight micronutrient fertilizers, while AAS method is used for analysis of various elements in micronutrient fertilizers mixtures because of low concentration and chances of interference form other elements.

Table-11: The reliable methods for various micronutrients are given below.

Element or Form	Volumetric method	Spectrophotometric method	AAS method
Zinc	EDTA titration method	Di- thiozone method	yes
Iron	$KMnO_4$ titration method	Orthophenonthroline method	yse
Manganese	-	Potassium periodate method	yse
Copper	$Na_2S_2O_3$ titration	Carbamates method	yes
Boron	NaOH direct titration method	-	Yes with N_2O flame
Molybdenum	-	Thiocynide $SnCl_2$ method	Yes with N_2O flame
Chelate Zn	EDTA direct and back titration method	-	-
Chelate Fe	EDTA direct and back titration method	-	yes

Preparation of sample of fertilizers:

- As per quantity given below in table 12 dissolve in water to prepare almost 1000 ppm solution. Dilute further to suitable standard curve range in table 13 with acidify water.

Reagents

1. Conc. HCl
2. Double distilled water (DDW)
3. Acidified water of pH 2.5 ±0.5; dissolve 10 ml of 10 % sulphuric acid in 10 litre of DDW and adjust the pH to 2.5 with the help of a pH meter by using drops of H_2SO_4 or NaOH.

Table-12: Sample solution preparation

Fertilizers	Quantity to be dissolve in litre water	Further dilution with acidified water
$ZnSO_4 . 7 H_2O$	0.25g	50 times
$ZnSO_4 . 1 H_2O$	0.25g	50 times
$FeSO_4 . 7 H_2O$	1.00g	20 times
Chelate Fe	1.00g	20 times
$MnSO_4 . H_2O$	0.40g	50 times
$CuSO_4 . 5 H_2O$	0.25g	100 times
Borax	10.00g	No dilution
Solubor	10.0g	No dilution
Ammonium molybdate	1.00g	10 times

Table-13: Working range of standard curve and wavelength

Element	Wavelength A°	Flame	Working range of standard curve(ppm)
Zn	2138	Air acetylene	0.2-2.0
Fe	2483	Air acetylene	2.0-20.0
Mn	2795	Air acetylene	0.5-5.0
Cu	3247	Air acetylene	0.2-1.0
B	2498	Acetylene N_2O	400-1600
Mo	3133	Acetylene N_2O	20-60

Preparation of standard elements solution

Prepare standard solution of 1000 ppm conc. of specific elements from specific standard materials as indicated below; further dilute it to suitable flaming range (Table 13).

Zinc: Dissolve 1 gm pure zinc metals in 30 ml of 6N HCl, heat to dissolve completely and dilute to 1000 ml. This is 1000 ppm Zn solution. Further dilute to the range of 0.2 to 2.0 ppm with acidified water.

Iron: Dissolve 1 gm pure zinc metals in 30 ml of 6N HCl, heat to dissolve completely and dilute to 1000 ml. This is 1000 ppm Zn solution. Further dilute to the range of 0.2 to 2.0 ppm with acidified water

Manganese: Dissolve 1 gm pure manganese metal in 30 ml of 6N HCl, heat to dissolve completely and dilute to 1000 ml. This is 1000 ppm Mn solution. Further dilute to the range of 0.5-5 ppm with acidified water.

Copper: Dissolve 1 gm pure Cu metals in 30 ml of (1+1 HNO_3), and heat to dissolve completely and dilute to 1000 ml. This is 1000 ppm Cu solution. Further dilute to the range of 0.2 to 1.0 ppm with acidified water.

Boron: Dissolve 44.095 gm of $Na_2B_4O_7.10\ H_2O$ in warm water, and dilute to 1000 ml to make 500ppm B solution. Further dilute to the range of 20-60 ppm with acidified water in 50 ml vol. flask

Molybdenum: Dissolve 1.50 gm MoO_3 in water, and dilute to 1000 ml. This is 1000 ppm Mo solution. Further dilute to the range of 20-60 ppm with acidified water.

Calculation:

$$\%\ \text{Micronutrient} = \frac{\text{X ppm} \times \text{dilution factor} \times 104}{\text{Wt of sample (g)}}$$

Where x is conc. of micronutrient in mg/kg(ppm) as read from curve.

Water analysis

Collection of Water Sampling Techniques: It is estimated that in India about 25% of ground water is of poor quality water. Water samples for irrigation quality assessment are to be analyzed for chemical constituents. Water sample from is collected in the bottle after raising it with the same water 3-4 times. Samples from a open wells and piezometers the sample bottle may be tied to a good quality thread if water from a tank or pond is to be tested, collected, the sample from at least 50 to 10 meters away from the boundaries. From a well the sample can be drawn either during irrigation just before the waterfalls the channel or by drawing it with the help of a bucket or any other clean container using a rope. From shallow ponds, reservoirs lakes, canals and rivers, water sample should be collected from centre of the width after avoiding the floating debris on the top. The water sample after proper marking and leveling must be sent to the laboratory immediately for testing to avoid any change or deterioration. If a few days delay is inevitable then 2 or 3 drops of pure toluene may be added to prevent bacterial activity. Proper labeling should be made on each sample as follows. i) Name ii) Location of body of water iii) Date iv) Time of sampling Depth of sampling. After labeling, immediately the samples must be sent to the laboratory for analysis to avoid any change or deterioration in its quality due to chemical and microbial activity. However, for appraisal of irrigation quality the water samples are mainly analyzed for total salts (EC) relative proportion of cations, anions and toxic substances such as excess boron and fluorine when sewage or any industrial effluent rich in organic matter is to be used for irrigation determination of Biochemical Oxygen Demand (BOD) and Chemical Oxygen Demand (COD) is also desirable.

Determination of pH:

Principle: The pH is a sort of voltage measurement and to cover the entire range 0- 14, a potential measurement in the range of +420 mv to – 420 mv is needed since a potential difference of 59.1 mv is developed for a difference of one pH unit.

Instrument: pH meter

Reagents: Buffer solution of pH 4.0 7.0 and 9.2

***Procedure*:** First standardize the pH meter with three pH buffer solutions after giving warm up times if needed. The pH is determined by taking about 50 ml water sample in a 100 ml clean beaker and immersing the electrodes of the pH meter in it. The pH is indicated on the dial, switch off the meter and remove and wash the electrodes.

Comments and suggestions:

- Never allow the lower portion of electrodes to touch bottom of the beaker.
- Ensure that the electrode is clean and filled with saturated KCl solution.
- The selector switch is set on mV.

36-Determination of Soluble salts

Principle: The total soluble salts can be determined by using conductivity meter or Solubridge as in case of soil water suspension which is expressed in terms of the sample's electrical conductivity (EC). The electrical conductivity is directly proportion to the ions and is expressed as dS /m or mmhos/ cm which is numerically same as micromhos/cm that is 1/1000th of deci Simons per meter or millimhos /cm.

EC or specific conductivity is the reciprocal of specific resistivity which is expressed as R α 1 /a, Rα p x 1 /a

Where R = electrical resistance (ohms)

1 = distance between electrodes, cm

a = cross section area, cm^2

p = specific resistivity ohms/ cm

Reagents: 0.01N KCl solution

Procedure: Switch on the conductivity meter and allow warm up time, if required and take the water sample in a clean beaker. Rinse the conductivity cell with distilled water and then with the sample. Standardize the EC meter with 0.01N KCl which gives a value of 1.41 dS /m at 25°C if the entire cell constant is 1.0. Dip the cell assembly in water sample taken in a 50 ml beaker and record the conductivity and adjust the current by rotating the position of the knob suitably.

Calculation:

Salt concentration (mg/L) = EC (dS/m) at 25 °C x 640

Salt concentration (me/L) = EC (dS/m) at 25 °C x 10

Osmotic pressure (atm) = EC (dS/m) x 0.36

Salt concentration in kg /ha.m.water = ppm x 10

me / 1 to mg / 1 (ppm) = multiply me / 1 of each ion by its equivalent weight and get the sum.

Values of EC usually vary by 2% per degree increase or decrease in temperature. Conductivity Increase with increase temperature

Determination of Anions: **Carbonates or bicarbonates**

Principal: *Carbonates* and bicarbonate in water can conveniently be determined by titration of a known volume of water against standard (usually 0.01N H_2SO_4) sulphuric acid using phenolphthalein and methyl orange indicators respectively. Addition of phenolphthalein gives red / pink colour in the presence of carbonates and titration with H_2SO_4 converts these CO_3^- and HCO_3^- and decolorize the red / pink colour. In colorless solution methyl orange is added which gives yellow colour. Thus, the carbonate neutralization is only half way. Ahead on complete neutralization of bicarbonates the yellow colour will change to red. Sometimes the bicarbonates titre value will be less if carbonate were not present due to absence of pink colour. In such a situation, either the same aliquot is used for bicarbonate titration or a fresh sample is analyzed for this. If carbonates are present and neutralized, the volume of H_2SO_4 used in the first phase (carbonate titration) is to be doubled to get the actual volume needed for complete neutralization of the carbonate.

Reagents:

- **Std H_2SO_4 (0.01N):** Take 2.8 ml of conc. H_2SO_4 to about 500 ml double distilled water (DDW) taken in a glass beaker and dilute to 1 L to get approx 0.1N H_2SO_4. Standardize the solution.
- **Phenolphthalein indicator:** Dissolve 0.25 g of phenolphthalein powder in 100 ml of 60 % ethyl alcohol.
- **Methyl orange indicator:** Dissolve 0.50 g of the dye in 100 ml of 95% ethyl alcohol.

***Procedure*:** Take a known volume of water sample in a clean porcelain dish and add about 20 ml of double distilled water(DDW) and add 2 drops of phenolphthalein indicator if pink/ red colour appears titrate against 0.01N H_2SO_4 till colour disappear. Now note volume of acid consumed. Now add 2 drop of methyl orange indicator in the colourless solution in the dish (left after the titration) or to a fresh taken 5 ml water sample. This develop a yellow colour and again titrate till the yellow colour changes into rosy red. Record the volume of acid consumed in the second titration.

Calculation:

$$\text{Carbonates (me / L)} = \frac{\text{BR x } H_2SO_4 \text{ normality} \times 1000}{\text{Sample (ml)}}$$

$$\text{Bicarbonates (me / L)} = \frac{(\text{FBR} - 2\text{ BR}) \text{ x } H_2SO_4 \text{ normality x } 1000}{\text{Sample (ml)}}$$

Here, BR is the burette reading (ml of H_2SO_4) after phenolphthalein is neutralized and FBR is final burette reading (total volume of H_2SO_4) after methyl orange. If no carbonates are present, FBR is written in place of (FBR-2 BR).

Note: The first titre value has been deducted from the second because the second volumes consumed for already present bicarbonates as well as for those obtained from neutralization of carbonates. The basis of the factor 2 is same as described for Ca + Mg in water and it has been double for carbonates because only half neutralization is done in the first titration.

***Determination of cations:* Calcium and Magnesium (Total Hardness):**

Principal The two main cations present in irrigation water sample are calcium and magnesium. For the determination of both of them together the versenate titration method is most popular used in which EDTA disodium salts solution is used to chelate them. The cations can also be estimated in water samples using AAS but for all practical purposes the versenate titration method is good enough. Calcium alone can also be estimated by versenate method using ammonium perpurate (Murexide) indicator and thus Mg can be obtained by deduction of Ca from Ca^{++} + Mg $^{++}$ content. Ca estimation can be done on a flame photometer but the precision is not very high. Based on the principle of cheletion the determination of CA+ Mg is carried out first by titrating the sample against EDTA after adding Eriochrome black –T (EBT) as indicator.

Reagent:

Std $CaCl_2$ solution (0.01N) : Dissolve exactly 0.5 g of AR grade $CaCO_3$ in minimum (10 ml) of 0.2 N HCl. Heat till the solution boils and CO_2 is completely expelled and then dilute to 1 L.

- **EDTA solution or versenate solution (0.01*N*):** Dissolve 2.0 g of ethylenediamine tetra acetic acid – dia sodium salts in water and adds 0.05 g of chemically pure magnesium chloride ($MgCl_2.6H_2O$) is added and dilute to 1 litre. Standardized it by titrating against 0.01N $CaCl_2$ solution.
- **Ammonium chloride –Ammonium hydroxide buffer (pH 10):** Dissolve 67.5 g of ammonium chloride in 570 ml of conc. ammonium hydroxide or ammonia solution, make the volume to 1 litre and adjust the pH at 10. **(Caution: ammonia bottle should be cooled before opening.)**
- **Sodium hydroxide (4 *N*):** 160 g of pure NaOH in water and make volume to 1 litre.
- **Ammonium perpurate (Murexide) powder:** Weigh 0.2 g of ammonium perpurate and 40 of potassium sulphate. Mix both the reagents thoroughly in a pestle mortal an keep it as powder in a clean dark colour bottle.
- **Eriochrome black–**T: weigh 0.5 g of EBT dye and 4.5 g of AR grade hydroxylamine hydrochloride in 100 ml of 95% ethanol.

Procedure: Take 5 ml of the water sample to a 100 ml porcelain dish and add about 20 ml of distilled water. Add 1 ml of buffer solution and 3-4 drops of EBT indicator and titrate with the std versenate solution (0.01N) or EDTA solution till colour changes to blue or bluish green. At the end point, no tinge of the red colour should remain in titrate sample. Rotate the dish to ensure complete disappearance of the red colour.

Calculation:

$$Ca^{++} + Mg^{++} \text{ me } / \text{ L} = \frac{\text{Volume of EDTA used (mL)} \times N \times 1000}{\text{mL of aliquot taken}}$$

$$Ca^{++} + Mg^{++}\ g\ /\ L = \frac{\text{Volume of EDTA used (mL)} \times N \times 32.196}{\text{mL of aliquot taken}}$$

Comments and suggestion

1. In Ca determination do not add more than 25 mg of ammonium perpurate otherwise colour identification is very difficulty.
2. For Ca + Mg, the indicator EDTA and buffer solution should be prepared fresh.
3. EDTA should be added drop by drop to get exact change of colour.
4. EDTA on storage settles down at the bottom. Before use either determine its normality again or prepare it fresh

Analysis of organic Constituents: Analysis or organic constituents in water and waste waters is generally carried out fewer than two categories. Organic substances which quantify an aggregates amount of organic matter. Important determination is of chemical oxygen demand (COD) indicating total organics present in the sample. Amount of total biodegradable organics in the water is indicated as biochemical oxygen demand (BOD).

Determination of Chemical Oxygen Demand (COD)

The chemical oxygen demand means the amount of oxygen required for oxidation of organic matter using a strong chemical oxidant such as potassium dichromate and potassium permanganate under reflux conditions. Potassium dichromate is the most suitable oxidant but for water having more than 2 g / l of chlorides, potassium permanganate is used. The test is widely used for estimate the degree of pollution in water and their self purification capacity, efficiency of treatment plants, pollution loads and a gives a rough idea of BOD which can be used for dilution of sample in BOD estimation. COD determination has an advantage over BOD test in that the results can be obtained in less than 5 hours whereas BOD determination requires 5 days. Further, the Cod test in relatively easy and precise. In addition, there is not much interference as in the case of BOD.

Principle: Decomposed organic matter produces CO_2 and H_2O when boiled with a mixture of potassium dichromate and sulphuric acid. A sample is refluxed with a known amount of potassium dichromate in sulphuric acid medium and the excess of dichromate is titrated against ferrous ammonium sulphate (FAS). The amount of dichromate consumed is proportional to the oxygen required to oxidize the oxidizable organic matter.

Apparatus: Reflux apparatus consisting of a flat bottom 250-500 ml capacity flask with ground glass joint and a condenser with 24 - 40 joint is needed. It is convenient to have a bigger hot plate on which a series of refluxing sets is connected to have a common water circulation to all.

Reagents:

- **Std potassium dichromate (00.25*N*):** dissolve 12.259 g potassium dichromate previously dry at 103°C in distilled water. Add 20 ml conc. H_2SO_4, cool and dilute to 1 litre.
- **Crystal:** Add 22 g mg silver sulfate to 4 g conc. H_2SO_4 bottle. Keep overnight and shake well after dissolution.
- **Std ferrous ammonium sulfate** (1N)*:* dissolve 98 g FAS in DDW and add 20 ml of sulphuric acid cool and dilute to 1 L. standardized it against std. $K_2Cr_2O_7$ and dilute 10 ml of this to about 100 ml. and add 30 ml conc. H_2SO_4 and cool. Add 3-4 drops of ferroin indicator and titrate with FAS till colour changes from green to a reddish blue or wine red.

$$\text{The normality of FAS} = \frac{\text{Volume of Potassium Dichromate (ml)} \times 0.25}{\text{Volume of FAS (ml)}}$$

- **Ferroin indicator:** Dissolve 1.485 gm of 1, 10- phenolphthalein monohydrate and 0.695 g or 695 mg ferrous sulphate ($FeSO_4.7H_2O$) in water and dilute to 100 L. Ready made indicator is also purchased form market.
- **Mercuric sulfate:** $HgSO_4$, crystal

Procedure: Place 0.4 g $HgSO_4$ in flux flax. Add 20 ml or an aliquot of sample diluted to 20 ml with DDW. Mix well so that chlorides are converted into poorly ionized mercuric chloride. Add 10 ml std. $K_2Cr_2O_7$ followed slowly by 30 ml sulphuric acid which already contains silver sulfate. This slow addition along with swirling prevents loss of volatile compounds such as fatty acids in the samples. Mix well. If the colour turns green either take a fresh sample with a smaller aliquot or adds more dichromate and acid. Final concentration of conc. H_2SO_4 should always be 50%. Connect the flask to condenser and flux for 2 hrs. Cool and wash down the condensers with small quantity of distilled water. Cool and titrate against std ferrous ammonium sulfate using ferroin indicator. Colour changes sharply from green blue to wine red. Reflux a reagent blank under identical conditions preferably simultaneously with the sample.

Calculation

$$\text{COD (mg/ l)} = \frac{(a - b)\ N \times 8000}{\text{Sample in ml}}$$

Where,

a = the ml FAS (Ferrous ammonium sulfate, Fe $(NH_4)_2$ $(SO_4)_2$ required for blank b = the ml FAS required for sample

N = normality of FAS (0.25), Equivalent of oxygen is 8

Comments and suggestions:

- The strength of sulphuric acid is in final solution should be 18N.
- The sequence of making mixture should be $HgSO_4$, sample (swirl), $K_2Cr_2O_7$ and conc. H_2SO_4 slowly with swirling.
- As far as possible, reflux blank and sample should be take simultaneously.
- Washing the condenser with DDW after fluxing and cool then titrate.

Determination of Biochemical Oxygen Demand (BOD): This term means the amount of oxygen required by the microorganism in stabilizing the biologically degradable organic matter under aerobic conditions. The determination of BOD involves the measurement of the oxygen conc. of the sample immediately and that after incubating it for 5 days.

Apparatus and Reagents

- **BOD bottles**
- **BOD incubator** having temperature control at 20°C
- **Phosphate buffer: dissolve** 8.5 g KH_2PO_4, 21.75 g K_2HPO_4, 33.4 g $Na_2HPO_4.7H_2O$ and 1.7 g NH_4Cl in distilled water to make up 1 Litre solution.
- **Magnesium sulphate solution** : 82.5 g $MgSO_4.7H_2O$ per litre
- **Calcium chloride solution :** 27.5 g anhydrous $CaCl_2$ per litre
- **Ferric chloride solution:** 0.25 g $FeCl_3.6\ H_2O$ per litre

Procedure

- Bubble compressed air in distilled water for 30 minutes this water is then called dilution water.
- In a litre of dilution water add 1 ml each of phosphate buffer $MgSO_4$, $CaCl_2$ and $FeCl_3$ solution and mix well.
- Neutralize the water sample to pH 7.
- The dissolved oxygen of the sample may not be sufficient for the oxidation of biologically degradable organic matter so that the sample is mixed with a known volume of dilution water which is rich in oxygen.
- Carry out dilution in a large trough or bucket and mix well.
- Fill two sets I and II of BOD bottles.
- Determine the dissolve oxygen in the sample bottles of set I immediately.
- Determine the dissolve oxygen in the sample bottles of set II after keeping the bottles in a BOD incubator at 20°C for 5 hrs.

Note: For blank determination take 2 BOD bottles and fill them with dilution water in once bottle determine the dissolve oxygen immediately and in the other, make the determination after incubation for 5 hrs.

Calculation

BOD in mg/ l = $(D_0\text{-}D_5)$ × dilution factor

Where Do = conc. of the dissolved oxygen in the sample

D5 = conc. of the dissolved oxygen in the sample after 5 days.

Biofertilizer

The term Biofertilizers has been coined to embody all such microorganism which add conserve and mobilize the plant nutrient in the soil. such microorganism have somehow come to be called as "biofertilizer" a term which is a misnomer, compared to commercial fertilizers manufactured on a large scale by factories.

In other words biofertilizers are based on renewable energy sources and are eco-friendly compared to commercial fertilizers. Biofertilizers play a very significant role in improving soil fertility by fixing atmospheric nitrogen, both in association with plant roots and without it solubilize insoluble soil phosphates and produce plant growth substances in the soil.

Classification

Depending upon the nutrient provided, Verma and Bhattacharyya (1994) have broadly classified biofertilizers as follows.

Biofertilizer (BF)

Nitrogen Fixing Biofertilizer (NBF)

NBF for legumes - Rhizobium

NBF for cereals - Azospirillum, *Azotobacter, Azolla, BGA*

Phosphate Mobilizing Biofertilizer (PMBF)

Phosphate solubilizer- *Bacillus Pseudomonas, Aspergillus*

Phosphate absorber - VA-Mycorriza (VAM) like Glomus

Soils of India: Soils are most important natural resources on which living beings are exiting from time immemorial. Soil are classified in order to study their properties with a view to increasing their productivity. India has a geographical area of 328,726 million hectares of the country and lies between latitudes 8° and 37° N and longitudes 69° and 93 °E. India has great diversity in physiography, climate, vegetable and soils derived from igneous, metamorphic and sedimentary rock.

Physiography: The major physiography divisions are (1) the northern mountains (ii) the great plains (iii) the peninsula and (iv) the island. The *northern mountains cover* the western Himalayas (Kashmir region, Himachal region, UP region) eastern Himalayas and Assam –Burma ranges. The

Great Plains consists of the Indus plains (Rajasthan and Punjab plains Ganga plains and the Bramaputra valley. The Peninsula consists of the central uplands, Deccan plateau (Satpura range, Maharashtra, plateau, Karnataka plateau Telengana plateau) eastern plateau (Bundelkhand plateau, Chhotanagpur plateau, Garhjat hills Mahanandi basin, Dandakaranya uplands) western hills (north Sahyadri, central Sahyadri Nilgiris, south Sahyadri eastern hill (eastern Ghats, Tamil Nadu upland) and coastal plains(west coastal plains, east coastal plains. The inlands small in land area, consist of the Arabian Sea islands and the Bay of Bengal Island. This broad region varies in their geological history, geological structure and physiographic character ataractics. The northern mountains extend all along the northern boundary of the country, bordering Pakistan on the west and Myanmar on the east. The Himalayas stretch out in a series of parallel or occasionally convergent chains separated by valleys or extensive platforms. The Peninsula is an ancient and stable land mass owing its present feature to denudation and weathering over long ages. The harder rock masses which have resisted weathering stands out as mountains and the softer ones from the valleys and plains. The whole terrain is built up of large and small undulating plateaus with their summits seldom rising to more than 1200 meters above means sea levels.

Climate: The climate of India is of monsoon type there being two distinct periods of rain fall in a year, *viz.*, the south west monsoon during the months of June to September and the north –east monsoon during the winter months. The climate is influenced by the Himalayan Mountains as well as the Indian Ocean, the Arabian Sea and the Bay of Bengal. The Himalayas obstruct the path of entry of cold winds from the north giving a continental type of climate. The seas produce a hot monosoonic type of tropical climate. The rainfall is controlled by topography as the high mountains lying across the path of monsoon winds help to precipitates their moisture as rain on the windward side. Therefore, the Western Ghats and the Assam ranges receive high rainfall (more than 200 cm per annual). The areas to the leeward side of these mountains receive comparatively little rain. As the south west monsoon is deflected towards upper India by the Himalayas, the Gangatic plains and Punjab and Haryana receive good rainfall. The moisture bearing winds from the Arabian Sea pass unobstructed over Rajasthan as the Aravallis lie along their path the rainfall in Rajasthan being scanty (around 25 cm per annum). the eastern parts of Peninsula receive some rain during the closing period of the south west monsoon and north east monsoon, the rain fall varying from 50 to 100 cm.

5
Introduction to Agrochemicals

Agrochemical or Agrichemical a concentration of agricultural chemical is a generic term for the various chemical products used in agriculture. In most cases, agrichemical refers to the broad ranges of pesticide including insecticides, herbicides, and fungicides. It may be also include synthetic fertilizers, hormones and other chemical growth agents and concentrated stores of raw animal manure. Many agrichemical are toxic and agrichemicals in bulk storage may pose significant environmental and or health risks, particularly in the event of accidental spills. In many countries, use of agrichemicals is highly regulated. Government issued permits for purchase and use of approved agrichemicals may be required. Significant penalties can results from misuse including improper storage resulting in spillage. On farms proper storage facilities and labeling, emergency clean up equipment and procedures and safety equipment and procedures for handling, application and disposal are often subject to mandatory standards and regulations. Usually the regulations are carried out through the registration process.

Pesticides are chemicals that are used in agriculture for control of pests, weeds and plant diseases. The chemicals natural and man-made have become an integral part of the human life. FAO defined pesticide as any substances or mixture of substances intended for preventing, destroying, or controlling any pest, including vectors of human or animal diseases unwanted species of plants or animals causing harm during or otherwise interfering with production, processing, storage, transport or marketing of food, agricultural commodities, wood and wood products, or stuffs, which may be administered to animals for the control of insect arachnids or other pests in or their bodies.

The words 'pesticides' is a very broad term. Literally it refers to killer of pests but it is also includes substances used for controlling, preventing, destroying, repelling or mitigating any pest. Pesticides can be classified in many different ways; according to target pest chemical structure of compounds used or the degree of type of health hazards involved. These chemicals may be extracted from plants or may be synthetic. Thus, the pesticides are of following categories.

Pesticides	Target
Insecticides	Substances that prevent destroy or kill insects
Fungicides	Substances that prevent destroy or inhibit growth of fungi ordisease in crops
Bactericides	Substances that prevent destroy or kill bacteria
Herbicides	Substances that prevent or inhibits growth of plants or kill weeds
Miticides	Substances that prevent destroy or kill mites
Rodenticides	Substances that prevent destroy or kill rodents
Nematocides	Substances that prevent destroy or kill nematodes
Chemo-sterilants	Substances that sterilize insects -pest
Molluscicides	Substances that prevent destroy or kill members belonging to phylum Mollusca or snail and slugs.
Plant growth	Substances that cause acceleration or retardation of rate of regulators growth or rate of maturation of plants
Defoliants	Substances that prevent plant leaves to die and fall away.
Desiccants	Substances that causing draining of moisture out of plantscausing them to dry.
Attractants	Substances that attract insect pest
Repellents	Substances that repel insect–pest from a treated plant
Avicides	Substances that kill or repels birds
Algaecides	Substances that kill algae
Larvicides	Substances that prevent destroy or kill larvae
Piscicides	Substances that prevent or kill fishes
Predicides	Substances that prevent or kill predators
Pheromone	Substances that attract insects
Disinfectants	Substances that prevent destroy or inactivate harmful microorganism

World over a large number of compounds have been synthesized showing insecticidal properties. In order to understand them, the insecticides can be classified on the following basis;

1. Based on toxicity,
2. Based on mode of entry
3. Based on mode of action, and

4. Based on chemical nature

1. **Classification based on toxicity:** Toxicity is the ability of the target animal. The toxicity is expressed in term of LD_{50} values. The dose in terms of mg of toxicant per kg of body weight (mg/ kg) that kill 50 % of the test animals.

   LD_{50} may be oral or dermal and respiratory toxicity (inhaled). The size of the dose is also important. On the application of single dose if the symptoms of acute stage of poisoning appear, it refers to acute toxicity. The accumulating effect of repeated doses is considered as chronic toxicity.

2. **Classification based on mode of entry:**
   - *(i)* **Stomach poison** – poison enters that the body of insect through food.
   - *(ii)* **Contact poison-** Poison enters the body of insect through cuticle.
   - *(iii)* **Fumigant-** Poison enters in gaseous state into the body of insects through spiracles and trachea.

3. **Classification based on mode of action:**
   - *(i)* **Physical poison:** Kills the insect by exerting the physical effect.
   - *(ii)* **Protoplasmic poison:** Kills the insects by destruction of cellular

Status of pesticides consumption in India

More than 2.6 million tonnes of active ingredients of pesticides are used world wide. Roughly 85% of this is consumed in agriculture. About three-quarters of pesticides are used in developed countries, mostly in Europe and Japan. India's pesticides are only 25 of the total world's consumption, but India is the third single largest consumers in the world and highest among the south Asian countries. During the last four decades consumption of pesticides in Indian has increased several hundred folds from 154 tonnes in 1953 -54 to 80,000 tonnes in 1994-95. However, their consumption steadily declined to 54,135 tonnes in 2000, primarily due to the ban on the use of Organochlorines pesticides such as HCH DDT Eldrin etc., which had high application rate. India is the largest producer of pesticides in south Asia; with production of 88,751 tonnes in 1999.

Acts and rules regulating use of pesticides in India

- The drugs and Cosmetics Act, 1940
- The prevention of food adulteration Act, 1954
- The prevention of food adulteration Act, 1955
- Insecticides Act, 1968
- Insecticides Rules, 1971
- Water (Prevention and Control of pollution) act, 1974
- Air (Prevention and Control of pollution) act, 1974

- Insecticides Order, 1986
- The environment Protection Act, 1986
- The consumer protection Act, 1981

At present there are 181 pesticides registered for use in agriculture and public health as given below. The pesticides can be banned or put under restriction for the use of the country by the central Insecticide Board to prevent harmful effects to environment and human health. At present 27 pesticides are banned and 7 are restricted use.

Insecticides Act, 1968

Import, manufacture, sale and distribution of pesticides is regulated under the Insecticides act,1968 and Insecticides Rules, 1971. There is a provision for registration of pesticides at the Central Government level and licensing for manufacturing and sale of pesticides by states / UT government after registration. The registration Committee constitute under section 5 of the Insecticides Act,1968 registers pesticides only after satisfying itself regarding their efficacy and safety to human beings and animals. If the pesticides are used as per the guidelines contained on their labels and leaflets, they do not cause any damage to human beings, animals or the environment.

Although the words "pesticides" includes insecticides fungicides, bactericides, rodenticides, nematicides and plant growth regulations etc, yet this legislation has been timed as "Insecticides Act" 1968. This act is not meant to regulate insecticides only but includes all types of pesticides including plant growth regulator. The insecticides Act 1968 is meant to regulate the import manufacture, sale, transport distribution and uses of pesticide with a view to prevent risk to human beings or animals and for matters connected there with. There were 38 sections of this act. As per the GSR 1650 exercising the powers conferred by section 36 of the Insecticides Act 1986 the central Government after consultation with the central Insecticides Board, the insecticides Rules 1971 were framed out having 46 rules. The Act provides for notification of four important functionaries for this purpose, viz., Licensing Officer Appellate Authority, Insecticides Inspector and analyst, to ensure that only genuine / legal action is taken against defaulter of law by the state / union Territories.

Further, the government has set up a Task Force in the department of Agriculture and Cooperation to get pesticide samples drawn for analysis to check quality thereof by the notification Central Insecticides Inspector.

There is network of 46 State Pesticide Testing Laboratories and 2 Regional Pesticide Testing Laboratories of the Central Government and the Central Insecticides Laboratory, established under Section 16 of the Insecticides Act, 1968 to test and analyze the quality of Insecticides. At the central level, Central Insecticides Laboratories has been set up under Section 16 of the Insecticides Act, 1968, to perform the statutory role of referral analysis.

Therefore, both central and state governments implement the Act. Under the Act, there are two statutory bodies,

One, the Central Insecticides Board, which is constituted under Section 4 of the Insecticides Act, 1968. The chairman of the Board is Director-General of Health Services, Ministry of Health and Family Welfare, Government of India. The function of the Board is to advise central and state governments on technical matter.

Pesticides registered in India under Insecticides Act.

2,4-dichlorophenoxy acetic acid	Barium carbonate	Carbosulfan
Acephate	Benfuracarb	Carboxin
Acetamiprid	Benomyl	Carta hydrochloride
Alachlor	Benthiocarb	Chlorimuron ethyl
Allethrin	Beta Cyfluthrin	Chlorimequ Chloride (CCC)
Alphacypermethrin	Bitertanol	Chlorofenvinphos
Alphanaphthyl acetic acid	Bromadiolone	Chlorpyriphos
Aluminium phosphide	Butachlor	Cinmethylene
Anilophos	Captafol	Clomazone
Atrazine	Captan	Copper oxychloride
Aureofungin	Carbarly	Copper Sulphate
Bacillus thuringiensis (B.t.)	Carbofuran	Coumachlor
Coumatetralyl	Fluchloralin	Paraquat dichloride
Cuprous Oxide	Flufenoxyuron	Penconazole
Cyfluthrin	Fluvalinate	Pendimethalin
Cyhalofop-butyl	Formothion	Permethrin
Cymoxanil	Fosetyl-Al	Phenthoate
Cypermethrin	Gibberellic acid	Phorate
Cyphenothrin	Glufosinate ammonium	Phosalone
Dalapon	Glyphosate	Phosphomidan
Dazomet	Hexacongenazole	Piroxofop-propanyl
Decamethrin/ Deltamethrin	Hydrogen Cyanamid	Clodinafop-propamyl
Diafethioron	Imazethapyr	Prallethrin
Diazinon	Imidaclorpid	Pretilachlor
Dichloro diphenyl	Imiprothrin	Primiphos-methyl
Trichloroethane DDT	Indoxacarb	Profenophos
Dichloropropane and	Iprodione	Propanil
Dichloropropane D	Dmixture Isoprothiolane	Propargite
Diclofop-methyl	Kasugamtcin	Propiconazole
Dicloros (DDVP)	Kitazin	Propineb

Contd...

Dicofol	Lambdacyhalothrin	Propoxur
Difenocenazole	Lime sulphur	Pyrethrins
Diflubenzuron	Lindane	Qunalphos
Dimethoate	Linuron	S-Bioallethrin
Dimethomorph	Malathion	Simazine
Dinicap	Mancozeb	Sirmate
Dithianon	Metalaxyl	Sodium cyanide
Diuron	Metaldehyde	Streptomycin+Tertacyclin
Dodine	Metasulfuron methyl	Sulfosulfuron
D-trans allethrin	Methabenzthiazuron	Sulphur
Edifenphos	Methomyl	Tebuconazole
Endosulfan	Methoxy ethyl mercury	Temephos
Eretilachlor	Chloride(MEMC)	Thiodicarb
Ethephon	Methyle bromide	Thiomethoxiain
Ethion	Methyl chlorophenoxy	Thiometon
Ethofenprox	acetic acid (MCPA)	Thiomethoxain
Ethoxysulfuron	Methyl parathion	Thiram
Ethylene dibromide and	Metolachlor	Transfluthrin
Carbon tetrachloride	Metoxuron	Triadimefon
Mixture(EDCT mixture)	Metribuzin	Triallate
Fenarimol	Monocrotophos	Triazophos
Fenazaquin	Myclobutanil	Tricholofon
Fenobucarb(BPMC)	Oxadiargyl	Tridemorph
Fenoxaprop-p-ethyl	Oxadiazon	Trifluralin
Fenfropathrin	Oxycarboxin	Validamycin
Fenthion	Oxydemeton-methyl	Warfarin
Fenvalerate	Oxyfluorfen	Zinc phosphide
Ferbam	Paclobutrazole	Zineb
Fipronil	Paradichlorobenzene(PDCB)	Ziram

Pesticides restricted for use in India

Aluminium phosphide	Methyl bromide	Methoxy-ethyl mercuric-DDT
	Methyl parathion	chloride (MEMC)
Lindane	Sodium cyanide	
Aldicarb	Chlorobromopropane	Nicotin sulphate
Aldrin	Copper acetoarsenite	Nitrofen
Benzene hexachloride	Dieldrin	Paraquate dimethyl sulphate
Calcium cyanide	Endrin	Pentachloro nitrobenzene
Carbofuran	Methomyl	Phosphamidon
Chlordane	Methomyl	Sodium methane arsonate

Another, statutory Body is Registration Committee (RC) constituted under Section 5 of the Insecticides Acts, 1968. Agriculture Commissioner, Government of India is the Chairmen of the Registration Committee. The function of RC is to register the pesticides after satisfying itself regarding efficacy of the pesticides and its safety to human beings, animals, and the environment. It also registers pesticide after scrutinizing the formulae verifying claims of efficacy and safety to human beings and animals specify the precautions against poising and any function incidental to these matter.

Pesticides refused registered

Azinphos ethyl	Dicrotophos	Mevinphos/phosdrin2,4,5-T
Azinphos methyl	EPM	Mephosfolan
Ammonium sulphamate	Fentin acetate	Thiodemeton / dislfoton
Binapacryl	Fenatin hydroxide	Vamidothion
Carnomethionate	Lead arsonate	
Chinomethionate/ morestan	Leptophos / phosvel	

* The list is subject to change from time to time.

Classification of pesticides based on chemical nature:

Inorganic Insecticides: Several mineral compounds of arsenic, fluorine, copper, mercury etc., which do not contain carbon in the composition, constitute inorganic insecticides. Examples- Aluminum phosphide, Calcium arsenate etc

Arsenicals: The arsenates are water insoluble and are safer to use plants than the water soluble arsenates which are phototoxic. They used only as poison baits are more rapid in their action than the arsenates.

Arsenites: It is white tasteless odourless compound about 20 times as toxic as DDT and is derivatives of arsenic trioxide (AS_2O_3). Paris green was the first commonly used arsenical developed for the control of Colorado beetle by 1865.

Arsenates: Compared to Arsenites, arsenates show less insecticidal activity but are safer to use on plants being highly water insoluble and hence non-phototoxic. Lead arsenate has very low pytotoxicity gives uniform deposits on plants and also does not decompose rapidly. Calcium arsenate is comparatively cheaper and more water-soluble than lead arsenate.

Fluorine Compounds: Fluorides are stomach as well as contact poison. Sodium fluoride (Na F) contains about 45% fluorine. The commercial product is required to be distinctly coloured to dispel. It is hazardous compound since 2-4 gram dose is fatal to humus. Sodium fluosilicate is highly toxic and also low in cost but has not been widely used due to its high pytotoxicity.

Cryolite (Sodium fluoaluminate, Na_3AlF_6) is one of the safest insecticides It can be used as 0.2% spray or as 4% dust.

Natural Occurring Insecticide or Bio-insecticide or Botanicals: Ground flowers, leaves or roots or their extractants were used for insect control. The toxic principles in botanicals are mostly alkaloids: heterocyclic nitrogenous compounds with marked physiological activity. Nicotine, Sabadilla, Ryania and pyrethrum are more important botanicals insecticides.

Natural products have enormous potentials as safe and biodegradable pesticides. The term biorational covers a range of such alternatives to synthetic chemicals their main feature is their increased specificity to avoid non-target mortality and other associated problem. The use of biorationals is an important component of our IPM strategy. These include plant products such as *Neem* or margosa, which have been shown to be effective against a number of pests; biopestiicdes such as nuclear polyhedrosis virus (NPV) for control of important pest like *Helicoverpa armigera;* and *Bacillus thuringiensis (Bt.)* which has gained importance in suppressing pests in crops like cotton and vegetables Commonly used botanicals are given in below.

Commercial neem –based Insecticides

Achook	Margosan-o	Nimbecidine	Phytowin
Azadit	Neemark	Nimbasal	Rakshak
Biosol	Neemzal-F	Nimlin	Repelin
Bioneem	Neemgold	Neem hit	Shaktiman
Econeem	Neemosan	Neemactine	Sukrina
Field marshal	Neemox	Neemocide	Sunnem
Jai neem	Nimba	Nimobas	Wellgro

Nicotinoids: Nicotine ($C_{10}H_{14}N_2$): Pure nicotine chemically known as 1-3-(1-methyl-2 pyrrolidyl) pyridine or 1-methyl -2,3 (pyridyl) pyrrolidine is a colourless odourless liquid soluble in water and most organic solvents. A commercial product of nicotine sulphate containing 40% alkaloids has been very popular for the control of aphids especially in vegetables @1-1.5 kg in 100 litre water for spray.

Nornicotine ($C_9H_{12}N_2$): It is resembles nicotine but the methyl group attached to pyrrolidine ring stands replaced by hydrogen atoms. Pure Nornicotine is a colourless, hygroscopic and slightly viscous liquid.

Pyrethrum:

Pyrethrum is known to occur in some species of Chrysanthemum. Pyrethrum insecticides are non toxic to warm-blooded animals and are also no phototoxic but they are photo-liable and unstable in alkaline medium. Cost is another factor which has also limited their use, now restricted only or the control of household pests. Pyrethrum are most toxic than cinerins.

Pyrethrums finds widest application in aerosols and oil based household sprays against various pests on account of excellent quick knock down properties they hold. Dust formulations have been used as grain Protestants. These have also successfully been used for the control of live stock ectoparasites and for limited pest control in vegetables.

Neem:

Neem, *Azadirachta indica* is grown through out India since time immemorial. It is very useful plant and utilized in diverse ways. Although all parts of the neem tree are biologically active but the highest activity could be seen in seed kernel. Neem seed kernels contains 0.2% -0.3% azdirachtion while leaves, seed coat and bark contains azadiractin to a lesser extant. Neem has been found to have a diverse effect on insects such as feeding deterrent growth disruptor, repellant, ovipositional deterrent insecticidal etc. The neem products are highly photo-degradable. The problem of pest resurgence has not been seen in neem treated crops. Neem seed extract at 12% conc. has been found as potential anti feedant for brown plant hopper (*Nilaparvata lugens*) by Saxena *et.al* (1980).Thus, neem is a very potent and has a great utility. Its use can be encouraged as it has no adverse effect on the environment and having no toxic residue in the harvested agro commodities. Consequently, a number of products are commercially available which are listed in the given table.

Some of the commercial available

S No	Neem products	Dose / ha
1.	Neem Mark	2.5 litre
2.	Nimebecidine	0.5 litre
3.	Raydon	1.25 to 2.50 litre
4.	Jawan	2.5 litre
5.	Nava Guard	1.0 litre
6.	Margosol	2.5 litre
7.	Crop Guard	2.5 litre
8.	Neemostar	2.5 litre
9.	Nemidin	1.5 litre
10.	Neem plus	1.0 litre
11.	Neemrich 20 EC	0.5 to 1.0 litre
12.	Azadit 20	1.0 litre

Organochlorines: These insecticides have carbon hydrogen and chlorine as their basic molecular constituents but oxygen and sulphur are also present in some insecticides belonging to this group. They are also known as chlorinated synthetics or or chlorinated organics or chlorinated hydrocarbons or chlorinated insecticides. The important groups of Organochlorines are:

- DDT and relatives
- Benzene derivatives
- Cyclodienes
- Chlorinated Tarpenes.

1. **DDT and Relatives:** DDT was first synthesized in 1873-74 by a German graduate student, Othnar Zeidler, who had no idea of its insecticidal activity. In 1939-40 a Swiss entomologist, Dr. Paul Muller, found its insecticidal properties for the first time. Its proved to be extremely effective against house flies and mosquitoes. This discovery could bring the noble Prize for medicine to Dr. Muller in 1948 for the life saving discovery. It belongs to the chemical class of diphenyl aliphatic

which means that it consists of an aliphatic or straight carbon chain with two phenyl ring. Chemically DDT is known as Dichloro Diphenyl Trichloethane. DDT is stomach and contact potent insecticide. It has got long residual action. It is also non-phototoxic except to cucurbits. It is being banned in several countries. It is stored in body fat and extracted in the milk. There are different related compounds of DDT which are still used as insecticides or acaricides. The examples are methoxychlor, dicofol, and chlorobenzillates ethylan.

2. **Benzene Derivatives:** Benzene (Six member carbon ring) is the main constituent of some insecticides. The first insecticides belonging to this group is BHC (benzene hexachloride. The true name of BHC is HCH or hexachlorocyclohexane. It was the first synthesized by Michael Faraday in the year 1825 but like DDT its insecticidal property was unknown till 1940. BHC is a stomach and contact insecticide. It has got slight fumigant action. It is persistent insecticide. It has been extensively used as soil insecticides particularly to control termites white grubs and cutworms. Among all these isomers only the gamma-isomer has the insecticidal activity. Now its use on agricultural crops has been banned by GOI. Highly purified product containing 99% of gamma isomer of HCH is also prepared and given name as Lindane
3. **Cyclodienes:** They are known as Cyclodienes because cyclo means cyclic or ring structure and diene means containing two double bonds. The Cyclodienes have three dimensional structures thus; possess stereo isomers whose atoms differ in their spatial location and structure. Cyclodienes also act as neurotoxicants. These insecticides were developed after world war-II. Some of the compound belonging to this group are Chlordane(1945) aldrin and dieldrin(1948) heptachlor (1949) eldrin (1951) endosulfan (1956) etc. Among them aldrin chlordane and heptachlor was often in use for termite control as they are most effective, long lasting and economical insecticides but now banned by GOI.
4. **Chlorinated Tarpenes:** There are mainly two insecticides namely toxaphene (1947) and storbane (1951) belong to chlorinated terpene group. Among these two, toxaphene is the insecticide which is considered as semi persistent materials as they are not persistent to the extent of Cyclodienes are. The mode of action of this insecticide is also similar to the Cyclodienes insecticides. They are act as on the neurons causing an imbalance in sodium and potassium ions.

Organophosphates

Those insecticides having phosphorus atoms come under the category of organophosphates. These insecticides can also be named as phosphate insecticides or organic phosphate etc. During Second World War II in germany Schrader and his co-workers in seeking substitutes for nicotine, found the insecticidal activity of an organic phosphate compound which led the development of several other insecticides belonging to this group. These insecticides have two most important properties such as higher potency and low residual life. These properties could bring them more popular day by day. Organophosphate compound are the combinations of different alcohols and different phosphorus acids. Thus they are the esters of phosphorus.

The organophosphates can be divided into three classes such as:

1. Aliphatic derivatives
2. Phenyl derivatives
3. heterocyclic derivatives

1. **Aliphatic derivatives:** The literary meaning of aliphatic means carbon chain and the linear arrangement of carbon atoms. Thus the phosphoric acid derivatives bearing short carbon chains come under this category such as Malathion, phorate dimethoate, demeton and monoctophos etc.
2. **Phenyl derivatives:** The insecticides belonging to this class contain a benzene ring with one of the ring hydrogen displaced by attachment to the phosphorus moiety and other frequently displaced by S, Cl, CH_3, CN, NO_2 etc. such as parathions tirofos and sulprofos etc.
3. **Heterocyclic Derivatives**: It indicates that the ring structure is composed of unlike atoms. In these compounds one or more carbon atoms is displaced by nitrogen, oxygen, sulphur and the ring may have 3.5 or 6 atoms. Such as chloropyriphos, diazinon, methidathion etc.

Phosphamidon ($C_{10}H_{19}ClNO_5P$)

The IUPAC name is 2-chloro-2-diethyl-carbamyl-1-methylvinyl dimethyl phosphate. In 1956 it was introduced by Ciba. The trade name is Dimecron. The insecticidal activity of Phosphamidon is mostly due to Z-isomer or -b-isomer. It is systemic insecticide, having low contact action. On application it is absorbed in the plant tissues within 1-3 hours. It is also non-phytotoxic.

Monocrotophos ($C_7H_{14}NO_5P$)

The IUPAC name is dimethyl -1-methyl-2-methylcarbamoylvinyl phosphate. It is systemic insecticides having contact action and widely used against a variety of insect pest from against pest soft bodies insects to leaf-eating beetles. At recommended dosages it is non-phytotoxic.

Parathion ($C_{10}H_{14}NO_5PS_2$)

The IUPAC name is *O, O-* diethyl *O-4-* nitrophynl phosphorothioate

(1) **Now its use in over country is banned.**

Chloropyriphos ($C_9H_{11}C_{13}NO_3PS$)

The chemical name is *O,O*-diethyl *O*-(3,5,6-trichloro-2- pyridyl) phosphorothioate. It is readily soluble in most organic solvents and soluble in water to the extract of 2 mg/l water. It can be rapidly detoxified in the animal body. It is non- systemic contact insecticides against household insects and crop pests.

Demeton ($C_8H_{19}O_3PS_2$)

The chemical name is demeton is approved by BSI and ISO for a reaction product comprising demeton–S. The trade name is systox. It is soluble to the extant of 60 mg/ l and 2 mg /l at room temperature.

Malathion ($C_{10}H_{19}O_6PS_2$)

The chemical name is S-1,2–bis (ethoxy-carbonyl) ethyl *O,O*- dimethyl phosphorodithioate. It is a contact insecticide. It has been recommended to control a variety of pests on crop. Since, it is safer and short lived insecticides. It is recommended on fruits and vegetables. It is mostly available in 5% dust and 50% EC.

Quinalphos ($C_{12}H_{15}N_2O_3PS$)

The IUPAC name of Quinalphos is *O,O*-diethyl O-quinoxalin -2-yl phosphorothioate. It is contact poison having good penetrating power. It can be degraded very fast in plants.

Phorate ($C_7H_{17}O_2PS_3$)

It's chemical name is *O,O*-ethylthiomethyl phosphorodithioate. It was introduced by the American Cyanamide Co. in 1954. It is formulated mainly as granules. The trade name is Thimet. It is very effective against sucking insect and also against maize borers, cut worm and white grubs etc.

Dimethoate ($C_5H_{12}NO_3PS_2$)

The chemical name is *O,O*-dimethyl S- methyl carbamoylmethyl phosphorodithioate. Its popular trade name is Rogor. It is most soluble in organic, solvents except paraffin. It is systemic insecticides widely used against sucking insect pests on various crops. It has also acaricides action.

Carbamates

The salts of carbamic acid are called carbaminates or carbamates. The chemists in European laboratories during 1863 found toxic alkaloid eserine in Calabar bean *Physostigma venonosum.* In 1864, the alkaloid was isolated, purified chemically identified and renamed as physostigmine. This was the first botanical carbamates and probably no other naturally occurring carbamates esters have even been reported.

Like organophosphate, the carbamates insecticides interfere in cholinergic transmission. The carbamate enters the synapse and inhibits the acetylcholine-esterase as a result the acetylcholine contains to depolarize the post synaptic membrane, causing prolonged stimulating resulting into the failure of the nerve or effectors tissue. The carbamates are of three type *viz.,* heterocyclic, phenyl and oxime carbamates.

Heterocyclic Carbamates

They are the esters of dimethyl carbamic acid. These are dimetan, isolan, pyrolan and primicarb etc.

Phenyl carbamates

N-Methyl carbamates are in widest use today. The first successful N-carbamate is the carbaryl. The details of few phenyl carbamates are given below.

Carbaryl ($C_{12} H_{11}NO_2$): This is the common name for the active ingredient 1-naphthyl N- methyl carbamate. It was introduced in 1996 by Union Carbide Corporation. The most popular trade name of carbaryl is Sevin. Carbaryl is produced by the reaction of 1-naphthol with methyl isocynate or with carbonyl chloride and methyl amine. It is most popular insecticides because it is effective against a wide range of insects and possesses vary low mammalian toxicity.

Carbofuran ($C_{12} H_{15}NO_3$): It was introduced by Bayer in 1967. It is a plant systemic broad spectrum and long residual insecticides. The popular trade name is Furadan. It is recommended as soil insecticides against plant sap sucking and borer pasts.

Landrin: It is a residual soil insecticide. It is composed of a 3:1 mixture of the 3, 4, 5- and 2, 3, 5 -isomer of trimethylphenyl N- methyl carbamate. It was introduced by Shell chemical company in 1968. It has been reported effectively against cutworms, wireworm, termite, white grubs etc.

Oxime carbamates: These are the esters of the oxime of aliphatic and cyclic aldehyde and ketenes.

Aldicarb ($C_7H_{14}N_2O_2S$): Its IUPAC name is 2-methyl-2-(methylthio) propionaldehyde, O-methycarbamoyloxime. It was introduced in 1965 by the union carbide corporation under trade mark 'Temik'. It is colourless crystal and soluble in water but insoluble in heptane. It is systemic pesticide usually applied in soil as seed furrow band or broadcast treatment either pre-plant or at planting as well as post emergence side dress treatments. Soil moisture is needed to release the active insecticide from the granule so irrigation is needed after application. It has also possessing acaricidal property.

Synthetic Parathyroids

The use of pyrethrum was not popularized in agriculture because of its high production cost, uncertainty of availability and instability in sunlight. In order to over come these difficulties, the synthesis of this pyrethrum like compounds came in vogue. Allethrin is probably first synthetic pyrethroid appeared in 1949. It is an ester made from alcohol having the ally side chain 2 –ally -4 hydroxy-3 methyl-2- cyclopentene-1-ones (allethriolone).After words several others synthetic pyrethroid such as cpermethrin, fluvalinate, decamethrin etc.,came into existence.

These synthetics pyrethroid are very less toxic to mammals and having a quick knock down activity to insects the lower toxicity to mammals and increase safety for the user. Very low application of synthetic pyrethroid as compared to conventional insecticides brings reduced environmental pollution. They have also a greater residual activity.

Cypermethrin: The IUPAC name is (RS)-a-cyno-3-pheoxybenzlyl-(RS)-cis,trans-3-(2,2 dichlorovinyl)-2,2-dimethyl cyclopropanecarboxylate. It was discovered by Elliott and was developed by Shell international chemical company. Ricord, Cymbush, imperator are same of the popular trade names. It is stomach as well as contact insecticides. It was very effective against different types of pest on various crops. The dosages usually recommended are ranging from 50 to 150 gram ai. / ha.

Fenvalerate: The IUPAC name is (RS)-a-cyno-3-pheoxybenzlyl-(RS)-2-(4-Chlorophenyl)-3- methyl butyrate. It is contact insecticides and broad spectrum in nature. It is marketed under the trade name Belmark, Sumicidin etc. The effective dosages against a wide variety of insect pest are ranging from 50 to 200 g ai / ha.

Biopesticdes

The synthetic insecticides described in earlier chapters are easily accessible to the farmers and are often used indiscriminately and unnecessarily leading to several hazards and environmental pollution. Efforts towards alternative insecticides are being made. Researches towards use of them are being packages. The details on neem have been described in earlier chapters. Besides, this there is a number of other insecticides which are biologically active, ecologically least disruptive and highly safe to man and animals are described under the head biopestiicdes whether they are originated from living or not.

Microbial Insecticides: The native pathogens reported on a host of insects include pathogen, virus, bacteria and fungi. Microbial insecticides are entomogenous above refereed pathogens and or their products that cause diseases in insect hosts, ultimately leading to death. There is relatively high degree of specificity in most cases which assist to protect non target organisms. They are harmless to men and animals

Viral pathogens: Insect virus belonging to the family Baculovirodae, based on the presence or absences of inclusion, site of multiplication either at nucleus or cytoplasm of cell, and shape are classified as:

(i) Nuclear Polyhedrosis Virus (NPV)
(ii) Cytoplasmic Polyhedrosis Virus (CPV)
(iii) Granulosis Virus (GV)

Nuclear Polyhedrosis Virus (NPV): The NPV is very effective against the larvae of Lepidoptera and to lesser extent to Hymenoptera and Diptera. The NPV acts as a true stomach poison. NPV is, therefore, extensively incorporated in IPM programme which is available from Pest Control India, Bangaloru.

Cytoplasmic Polyhedrosis Virus (CPV)	:	They are slow acting and retard the normal growth of insect.
Granulosis Virus (GV)	:	Next to NPV, the GV is important.

Bacterial Pathogens:

Bacillus thuringiensis (B.T.) : Among the various spore forming bacterial pathogens gram positive, rod shaped, spore forming and crystalliferous bacteria *Bacillus thuringiensis* commonly called B.t. is one of the most effectively and extensively used entomobocterial pathogens for control of almost all lepidopterous pests. It was detected in 1902 in dying larvae of Bombyx mori by S. Ishiwata and was first studied and named after isolation from larvae of *Ephestia kuehniella* by Berliner in 1915 in Germany. It was first used as a microbial insecticide 'sporeine' against several pesticdal companies the world over.

Synergism of insecticides: Synergism may be defined as an increase in the bioactivity of two compounds to produce an effect greater than the one expected from a single algebraic summation of the effects of two compounds individually administered.

Thus synergist may be defined as any chemical which in itself is not toxic to insects as dosages used, but when combined with an insecticide, greatly enhances the toxicity of the insecticides If two materials like one toxic material and other modifier applied together elicit lower response than that expected from the sum of two individuals is known as the antatogonism. This is also known as negative synergism.

Some of the synergists used with insecticides are sesamin, seasamolin, piperonyl butoxide, semax, safrole, sulfoxide, propylison tropital.etc. The most useful synergist is the methlenedioxyphenyl (mdp) moity.Antagonistic action has been observed when piperonyl cyclonene was used as synergist with malathion, methyl parathion and diazinon. Antagonistic action also appears if parathion is mixed with malathion parathion with azinphosmethyl, dematon with trichlorfon.

Synergism and antagonism can mathematically be expressed calculating co-toxicity coefficient based on LC_{50} values determined for individual insecticide and the synergist or non- toxic material mixed insecticide

$$C.C = \frac{I}{M}$$

Where,

CC = co-toxicity coefficient,

I = LC_{50} of insecticides and synergist against test Insect,

M = LC_{50} of mixture of insecticides and synergist against test insect.

If $CC > 1$ = Synergism

$CC < 1$ = Antagonism

6

Objective Questions

MULTIPLE CHOICE

1. Illite, Montmrillonite and Kaolinite are:

(a) Kinds of soil microbes (b) Kinds of soil in USA

(c) Kinds of protozoa in soil (d) Clay minerals

2. A vertical section cut downwards through the soil is known as:

(a) Soil profile (b) Soil samples

(c) Soil column (d) Parent material

3. A branch of science that deals with the study of origin of the soil its classification and its description is known as:

(a) Pedology (b) Edaphology

(c) Ecology (d) Geology

4. Soil Horizon A, E and B together is:

(a) Soil solum (b) Epipedon

(c) Soil profile (d) None of the above

5. **The arrangement of the different soil aggregates in a particular pattern is known as**
   (a) Soil solum (b) Epipedon
   (c) Soil structure (d) Soil horizon
6. **The concept that mineral elements in the soil and in added manures and fertilizers are essential for plant growth was given by**
   (a) Kjeldhal in 1973 (b) Julius Von Leigib in 1840
   (c) Woromin in 1972 (d) Baisingault in 1793
7. **Nitrogen fixing bacterium called *Azotobacter Flourishes* is developed under:**
   (a) Legumes (b) Teak plantation
   (c) Grass land (d) Pine forest
8. **The pyrite is mostly found in**
   (a) Bihar (b) Rajasthan
   (c) AP (d) Maharastra
9. **Under high saline irrigation water we will prefer to grow**
   (a) Pea (b) Gram
   (c) Wheat (d) Barley
10. **Zinc Sulphate is generally added to take more production under:**
    (a) Acid soils (b) Water logged soil
    (c) Wheat (d) Alkaline soil
11. **Available Zn in the soil, can be determined by using the solution of**
    (a) 1N NH_4 OAC (b) EDTA + $(NH_4)2\ CO_2$
    (c) DTPA + $CaCl_2$(pH7.3) (d) All of these
12. **Olsen extractant is used for available P determination in**
    (a) Neutral to alkaline soil (b) Arid soil
    (c) Acidic soil (d) Waterlogged soil
13. **Diphenyl amine is an indicator which is used in**
    (a) O C (b) Phosphorus
    (c) Potassium (d) Fe
14. **Irrigation with sodic water increase the availability of**
    (a) Zn (b) Na
    (c) Mo (d) Cl
15. **The sodicity of water can be expressed by**
    (a) ECe (b) ESP
    (c) SAR (d) None of these

16. Biological nitrogen fixation is

(a) Both aerobic and anaerobic process
(b) aerobic process
(c) anaerobic process
(d) None of these

17. De-nitrification is more in

(a) Well drain soil (b) Waterlogged soil
(c) Heavy soil (d) Light soil

18. Till cake content about –N%

(a) 6.0 (b) 10.2
(c) 9.8 (d) 2.5

19. In India, the area under red soil is (m.ha)

(a) 8.0 (b) 5.1
(c) 7.5 (d) 10.5

20. Horizontal layer in the soil profile are referred to as

(a) Horizontal zones (b) Rhizosphere
(c) Soil horizons (d) Soil column

21. Ammonia is transferred into nitrate by

(a) Fungi (b) Algae
(c) Bacteria (d) Actinomycetes

22. Soil is biochemically withered part of

(a) Parent Material (b) Organic Matter
(c) Regolith (d) Bedrock

23. A mass of unconsolidated debris varying in composition from place to place and covering bedrock is defined as

(a) Soil horizon (b) Soil debris
(c) Regolith (d) Soil profile

24. Downward movement of water is controlled by:

(a) Soil pH (b) Soil temperature
(c) Texture of parent material (d) Nature of soil microbes

25. Quartzite are metamorphic rocks derived from

(a) Slate (b) Granite
(c) Limestone (d) Quartz

26. Alluvial soil in India are:

(a) Very rare (b) Most extensive soils
(c) Found in desert (d) Least fertile

27. Hydraulic conductivity of soils varies directly with
(a) Total porosity (b) WHC
(c) Capillary porosity (d) Non - Capillary porosity

28. Phosphorus fixation refers to the phenomenon of
(a) Conversion of soluble phosphorus into plant unavailable forms
(b) Absorption of phosphorus by the plant and synthesis biological compounds
(c) Leaching of phosphorus by percolating waters
(d) Anion exchange with colloidal constitute of soil

29. The area of black soil in India (M.ha)
(a) 5.46 (b) 75.0
(c) 54.6 (d) 28.5

30. The area of Hill soils in India (M.ha)
(a) 5.46 (b) 1.3
(c) 54.6 (d) 28.5

31. The black soil in the largest group of Indian soils
(a) First (b) Second
(c) Third (d) None of these

32. Which state has the maximum black soil
(a) Haryana (b) Bihar
(c) Maharastra (d) Uttar Pradesh

33. Black soil are ————when wet and very hard when dry
(a) Hard (b) Very soft
(c) Very hard (d) Soft

34. Black soils have —— amount of lime and P
(a) High (b) Low
(c) Very high (d) Very low

35. The elements play direct role in formation of chlorophyll are
(a) P and K (b) N and Mg
(c) Ca and Zn (d) Fe and S

36. The major nutrients are:
(a) Fe, Mn, Zn (b) N, P, K
(c) Cu, Mo, B (d) Co, Cl, Na

37. Khaira disease of rice is due to the deficiency of
(a) Fe (b) Zn
(c) N (d) ca

38. Amount of heat required to raise temperature of 1 g soil from 14 to 15 ^{o}C is-

(a) Thermal conductivity (b) Heat capacity
(c) specific heat (d) None of these

39. Soil structure units having horizontal axis much longer than vertical axis is-

(a) Platy (b) Blocky
(c) Prismatic (d) Columnar

40. The arrangement of the different soil aggregates in a particular pattern is-

(a) Soil structure (b) Soil texture
(c) Soil consistence (d) Soil dispersion

41. At field capacity moisture is held at

(a) 31 atm (b) 15 atm
(c) 1/3 atm (d) None of these

42. Specific heat of the dry soil is equal to (al./g)

(a) 0.2 (b) 1.0
(c) 0.5 (d) None of these

43. Gravitational potential in soil water is potential created due to

(a) Mass of the soil solids (b) Presence of organic matter
(c) Soil solution (d) Gravity

44. Which elements are mobile in soil

(a) N (b) Mg
(c) K (d) Ca

45. If the particles size is 2.0 mm to 7.5 cm than it is known as

(a) Silt (b) Stones
(c) Gravels (d) Sand

46. By the formula d= P w X As X D/100 we calculate

(a) Quality of irrigation water (b) Quality of irrigation
(c) Depth of irrigation water (d) none of these

47. Lithosphere has about —%——Igneous rocks

(a) 75 (b) 95
(c) 80 (d) 72

48. Lithosphere contain about——%— Shales

(a) 9 (b) 1
(c) 4 (d) 0.75

49. Which is not a Igneous rocks
(a) Granite (b) Dibase
(c) Quartzite (d) Marble

50. The new comprehensive soil classification is based on:
(a) Physical and chemical properties of soil
(b) Genetic characters
(c) Both (a) and (b)
(d) None of these

51. 'A' horizon of the soil is also designated as
(a) Eluviated horizon (b) Illuviated horizon
(c) (A) and (C) both (d) None of these

52. The characteristics of saline soil is
(a) ECe<4dS/m,pH<8.5,ESP>15
(b) ECe>4dS/m,pH<8.5,ESP<15
(c) ECe <4dS/m,pH<8.5,ESP<15
(d) ECe<4dS/m,pH>8.5,ESP>15

53. Lime is added to reclaim:
(a) Alkali soil (b) Acidic soil
(c) Saline soil (d) Laterite soil

54. Poor quality waters contains
(a) Higher amount of soluble salts
(b) Higher amount of soluble Ca
(c) Lower amount of soluble salts
(d) None of these

55. Alkali soil are reclaimed by the addition of :
(a) Gypsum (b) Lime
(c) Calcium Chloride (d) All of these_

56. Which is not a metamorphic rock
(a) Marble (b) Quartzite
(c) Schists (d) Sandstone

57. Kaolinite is a ———type clay mineral
(a) 2::1 (b) 2::2
(c) 1::1 (d) None of these

58. $Si_4A_{10}(OH)_8$ is a formula of
(a) Muscovite (b) Kaolinite
(c) Mica (d) Biotite

59. C-Spacing of Kaolinite is (^{o}A)

(a) 7.39 (b) 14.0

(c) 7.93 (d) None of these

60. $Si_4O_{10}(OH)_8$ n H_2O is a formula of

(a) Kaolinite (b) Muscovite

(c) Halloysite (d) Vermiculite

61. $K_2(Si_5Al_2)Al_4O_{20}(OH)_4$ is the formula of

(a) Muscovite (b) Biotite

(c) Kaolinite (d) Montmrillonite

62. The CEC of Mica is about (cmol/100 g)

(a) 50 (b) 140

(c) 10 (d) 100

63. The CEC of Illite is about (cmol/100 g)

(a) 40 (b) 10

(c) 140 (d) 100

64. Montmrillonite and Vermiculite are the type of mineral

(a) 1::2 (b) 1::1

(c) 2::1 (d) None of these

65. Terai soils are deficient in

(a) Zn (b) N

(c) S (d) P

66. Slates are metamorphic rocks derived from

(a) Lime stone (b) Shales

(c) Basalt (d) Granite

67. Change in texture and mineral composition of Igneous and Sedimentary rocks by the action of intense heat and high pressure results in the formation of

(a) Compact rocks (b) Resistance rocks

(c) Metamorphic rocks (d) Decomposable rocks

68. Rocks from which the soils develop are called

(a) Parents soils (b) Soils materials

(c) Parents materials (d) Primitive soils

69. Soils containing high proportion of organic matter have the colour

(a) Dark (b) Red

(c) Yellow (d) Light

70. Alluvial soils of UP are generally

(a) Flooded (b) Under forests

(c) Dry (d) Under pastures

71. Conglomerate, Sandstone Shale and Lime stone are
(a) Metamorphic rocks (b) Igneous rocks
(c) Sedimentary rock (d) Morphogenic rocks

72. The common Igneous rocks found in India are
(a) Sedimentary rock (b) Morphogenic rock
(c) Granite, Basalts or Deccan trap
(d) Conglomerate, Shales

73. In India black soils are the
(a) Largest group (b) Least fertile
(c) Second largest group (d) Smallest group

74. Saline soils occur mostly in
(a) Punjab (b) J & K
(c) MP (d) Haryana

75. Gneiss in metamorphic rocks derived from
(a) Granite (b) Slate
(c) Lime stone (d) Quartz

76. The chemical decomposition of rocks is brought about by
(a) Solution and Hydrolysis (b) Oxidation and Reduction
(c) Carbonation (d) All of these

77. The common sedimentary rocks are
(a) Conglomerate (b) Shale and lime stone
(c) Sandstone (d) All of these

78. Phosphorous Fixation refers to the phenomenon of:
(a) Absorption of P_2O_5 by the plant and synthesis of biological compound
(b) Conversion of soluble phosphorous into plant unavailable forms
(c) Anion exchange with colloidal constituents of soil
(d) Leaching from soil by percolating waters

79. The physical agents of physical weathering are:
(a) Temperature (b) Plants and animals
(c) Wind and water (d) All of the above

80. How many kinds of main rocks are there
(a) 2 (b) 3
(c) 1 (d) 4

81. The components of soil are
(a) Mineral material (b) Water and Air
(c) Organic matter (d) All of these

82. The common igneous rocks found in India are-
(a) Basalt of Deccan trap (b) Granite
(c) None of the both (d) Both (a) and (b)

83. The CEC value of Montmrillonite is ———meq/100 gm soil or cmol/ 100 kg soil
(a) 80-140 (b) 80-160
(c) 80-150 (d) 80-170

84. Organic carbon is found ————% in deep black soil-
(a) 0.34-0.77 (b) 0.068-6.53
(c) 0.28-1.10 (d) None of these

85. A unit of illite consists of -
(a) Potassium (b) Alumina
(c) Silica (d) All of the above

86. Relative proportion of soil separates in sand silt and clay is termed as-
(a) Soil texture (b) Soil structure
(c) Soil porosity (d) None of these

87. The alkali soil have an excess of
(a) Calcium (b) Sodium
(c) Zinc (d) Potassium

88. The alkali soil can be identified by seeing
(a) Red colour on the surface
(b) White layer of salts on the surface
(c) Black colour on the surface
(d) None of these

89. The salt tolerant crop is
(a) Pea (b) Barley
(c) Loamy soil (d) None of these

90. Alkali soil are generally found in—
(a) Sub-humid climate (b) Arid and semi-arid climate
(c) Humid climate (d) Temperate climate

91. Which is the submicroscopic agents
(a) Bacteria (b) Fungi
(c) Viruses (d) Actinomycetes

92. Fungal population is comparatively high in a soil having
(a) Alkaline pH (b) Acidic to neutral pH
(c) Acidic pH (d) Neutral pH

93. Nitrification in the soil takes place by
(a) Oxidation process (b) Biological oxidation
(c) Reduction process (d) Both (a) & (b)

94. Dominant organisms found in a compost pits are
(a) Thermophilic (b) Psychrophilic
(c) Mesophylic (d) Both (b) & (c)

95. The salt tolerant crop is
(a) Pea (b) Barley
(c) Loamy soil (d) None of these

96. Organisms used as biofertilizer in wheat cultivation
(a) *Azotobacter* spp. (b) *Rhizobium* spp.
(c) Both (a) & (b) (d) None of these

97. Urea contain
(a) 24% N (b) 16% N
(c) 60% N (d) 46% N

98. If the soil has more than 15% exchangeable sodium it is considered to be
(a) Neutral (b) Acidic
(c) Alkali (d) None of the above

99. A unit of Kaolinite crystal lattice of silica and aluminium in the ratio of-
(a) 1::2 (b) 1::3
(c) 2::1 (d) 1::1

100. The ratio of organic carbon to organic matter in soil is-
(a) 1.0::1.7 (b) 20.0::1.0
(c) 1.0::2.0 (d) 1.7::1.0

101. Humus soil contains C,O_2,N,H_2 and ash in the ratio of -
(a) 10-7-1-1-1 (b) 1-7-10-1-1
(c) 7-10-1-1-1 (d) None of the above

102. Pedology means-
(a) A study of crop husbandry (b) A study of soil profile
(c) A coarse grained soil (d) None of the above

103. Humus soil contains carbon %
(a) 0.5 (b) 35
(c) 50 (d) 5.0_

104. The CEC value of illite is ——cmol/100 g soil
(a) 1-30 (b) 10-20
(c) 10-40 (d) 20-30

105. The CEC value of kaolinite is ——cmol/100 kg soil-

(a) 3-10 (b) 80-150
(c) 10-30 (d) None of the above

106. Which element is considered as energy currency for the plant-

(a) N (b) P
(c) K (d) Mo

107. Which element is indirectly related with drought resistance -

(a) N (b) P
(c) K (d) Zn

108. The purpling along the leaf edges is the characteristic deficiency symptoms -

(a) P (b) K
(c) B (d) Fe

109. Which one of the following element is considered to be an immobile element in the plant

(a) Ca (b) Mg
(c) N (d) P

110. The soil is well flocculated if it has more of-

(a) Na (b) K
(c) Zn (d) Ca

111. Complete chlorosis of new leaves is generally due to the deficiency of nutrients-

(a) Zn (b) Mn
(c) Fe (d) Cu

112. Rock phosphate is used in

(a) Alkaline soil
(b) Acidic soil
(c) Neutral soil
(d) Saline soil

113. In saline-alkaline soil the electrical conductivity is- (d S /m)

(a) 2 (b) 3
(c) >4 (d) None of these

114. Soils with high pH are generally deficient in

(a) Zn and Mn (b) Ca and Mg
(c) B and Fe (d) Cu and Mo

115. Which one of the following minerals is least weathering in soils-

(a) Olivine (b) Biotite
(c) Quartz (d) Calcite

116. C::N ratio of organic matter is -

(a) 10 to 1 (b) 30 to 2

(c) 20 to 2 (d) None of these

117. A soil with 20meq /100 g CEC and 5meq/100 g exchangeable H will have a base saturation of

(a) 80 % (b) 25 %

(c) 75 % (d) 133.3 %

118. N P and humus are generally poor in -

(a) Laterite soil (b) Alluvial soil

(c) Cotton soil (d) Red soil

119. Which of the following fertilizer has maximum plant nutrient-

(a) Muriate of potash (b) DAP

(c) TSP (d) Ammonium nitrate

120. Fe deficiency is mostly seen in crops growing on

(a) Calcareous or alkaline soils

(b) Salt affected soil

(c) Acidic soil

(d) Neutral soil

121. Fe is important component of enzymes

(a) Nitrogenase (b) Nitrate reductase

(c) Nitrite reductase (d) All of these

122. Bronzing is associated with Fe levels greater than 300 ppm in the leaf blade of

(a) Wheat (b) Rice

(c) Maize (d) Sorghum

123. Crinkle leaf of cotton is caused by the toxicity of -

(a) Fe (b) Cu

(c) Mn (d) Zn

124. Mo is a component of -

(a) Urease (b) Nitrogenase

(c) Nitrate reductase (d) Both (b) and (c)

125. Fertilizers containing Cl^- ions are not suitable for-

(a) Sugarcane (b) Potato

(c) Wheat (d) Pigeon Pea

126. Intercellular deposits of which element is known as plant opals

(a) Na (b) Si

(c) Ni (d) V

127. Which element is the metal component of urease enzymes -

(a) Si (b) Na

(c) Mo (d) Ni

128. The negative charge on 2::1 clays is-

(a) pH dependent (b) pH independent

(c) Both (a) and (b) (d) None of these

129. Ammonium sulphate contain —N %

(a) 49 (b) 20

(c) 42 (d) 46

130. What is the P_2O_5 content of various rock phosphate -

(a) 20-35 % (b) 49%

(c) 11-20% (d) 18-20%

131. The fertilizers which have two or more than two major plant nutrient in their chemical composition are called -

(a) Complex fertilizer (b) Organic fertilizer

(c) Double fertilizer (d) Mixed fertilizer

132. For nitrification, the optimum pH is -

(a) 6.5 (b) 7.5

(c) 8.5 (d) 9.5

133. *Nitrosomonas and Nitrobacter* cause nitrification are an obligate -

(a) Autotrophic bacteria (b) Heterotrophic bacteria

(c) Both (a) and (b) (d) Saprophyte

134. Maximum nitrification takes place at——% of oxygen

(a) 10 (b) 20

(c) 30 (d) 40

135. Permissible biuret level in most of urea fertilizer is ————%-

(a) 4 (b) 2

(c) 3 (d) 5

136. NH_4^+ fixation takes place in

(a) Vermiculite and Illite (b) Montmrillonite and Kaolinite

(c) Chlorite and Kaolinite (d) All of these

137. Which fertilizer has risk of fire or even explosion -

(a) Aqua NH_3 (b) Ammonium nitrate

(c) Potassium nitrate (d) Ammonium chloride

138. Biuret in urea applied as foliar spray in citrus fruits should not be more than in-—%

(a) 0.25 (b) 0.50

(c) 0.75 (d) 1.50

139. Which fertilizer act as a nitrification inhibitor and as a slow release N source -

(a) CDU (b) IBDU

(c) N-serve (d) DCD

140. At what pH, there are approx. equal amounts of $H_2PO_4^-$ and HPO_3^{2-}

(a) 6.5 (b) 7.0

(c) 7.2 (d) 8.5

141. The inorganic and organic labile P fraction of soil are collectively called -

(a) Intensity factor (b) Quantity factor

(c) Capacity factor (d) None of these

142. 'Peat' soil are generally characteristics as

(a) Black, heavy and acidic

(b) Red, light and acidic soil

(c) Black, light and alkaline soil

(d) Black, heavy and neutral soil

143. P content in DAP is ——%.

(a) 20 (b) 36

(c) 46 (d) 50

144. What is the porosity percentage if the particle density of soil is 2.6 and bulk density is 1.3 Mg/m

(a) 35% (b) 40%

(c) 50% (d) 55%

145. Height of rise of water in a capillary tube is directly proportion to

(a) Surface tension of water (b) Radius of capillary tube

(c) Viscosity of water (d) Both (a) and (c)

146. Hill soils are generally

(a) Acidic (b) Alkaline

(c) Neutral (d) Saline

147. The diagnostic surface horizons are called

(a) Genetic horizon (b) Pedons

(c) Epipedon (d) Endopedons

148. Available water holding capacity is maximum in

(a) Clay loam (b) Sandy clay loam

(c) Silty clay loam (d) Loam

149. Soil aeration limits plants growth when % pores filled with water are more than

(a) 50-60% (b) 60-70%

(c) 70-80% (d) 80-90%

150. Soil colour determination, Chroma refers to which of the following.
(a) Dominant spectral colour
(b) Lightness or darkness of colour
(c) Purity or strength of colour
(d) None of these

151. Chemical formula of pyrites reclaiment is
(a) Fe S (b) Cu S
(c) Mn S (d) FeS_2

152. Concentration of which element is highest in soil
(a) Fe (b) O_2
(c) Al (d) Si

153. Which is an incorrect match in the following?
(a) Soil bulk density :: Hydrometer
(b) Particle size analysis of soil :: Picnometer
(c) Soil moisture retention :: Permeameter
(d) Soil moisture estimation :: Potometer

154. Specific heat is higher in
(a) Sand (b) Silt
(c) Clay (d) Humus

155. Laterization is a process of accumulation in soil-
(a) Iron and aluminium (b) Silica
(c) $CaCO_3$ (d) $CaSO_4$

156. Podzolization is a process of accumulation
(a) Silica (b) Fe and Al
(c) $CaCO_3$ (d) $CaSO_4$

157. Rocks containing more than 65% SiO_2
(a) Neutral (b) Alkaline
(c) Acidic (d) Calcareous

158. Average content of silica in soil is
(a) 24% (b) 27%
(c) 31% (d) 35%

159. Which is a incorrect match in the following
(a) Kaolinite :: P (b) Illite :: Iron
(c) Montmrillonite :: K, Mg (d) Vermiculite :: Mg

160. Critical limits of available Phosphorus in soil for wheat crop by Olsen's method is – kg/ha
(a) 5 (b) 10
(c) 15 (d) 22.5

161. Sulphur content in ammonium sulphate is

(a) 12% (b) 20%

(c) 24% (d) 30%

162. Which one is dominating salt in alkali soil

(a) $CaCl_2$ (b) $MgSO_4$

(c) $NaHCO_3$ (d) Na_2CO_3

163. In which soil zinc availability to crop is higher-

(a) Acidic soil (b) Alkaline soil

(c) Calcareous soil (d) Waterlogged soil

164. Which of the following is correct

(a) P_2O_5 X 0.44 = P (b) P_2O_5 X 2.29 = P

(c) P_2O_5 x 2.29 = P (d) P_2O_5 x 1.20 = P

165. The major organic cementing agent in soil aggregate formation is

(a) Lipids (b) Protein and protein derivatives

(c) Polysaccharides (d) Organic acids

166. Amongst oil cake, the highest nitrogen content is in-

(a) Castor cake (b) Neem cake

(c) Coconut cake (d) Groundnut cake

167. The element present in the porphyrin structure of chlorophyll molecules are-

(a) Mg (b) Fe

(c) N (d) Mg and N

168. Which one of the following is a manganese containing minerals-

(a) Magnetite (b) Tourmaline

(c) Apatite (d) Pyrolusite

169. Plasticity of soil can be determined by using

(a) Pycnometer (b) Buoyoucos hydrometer

(c) Atterberg's apparatus (d) IRG

170. Soil aggregation can be estimated by

(a) Yoder's apparatus (b) Cone penetrometer

(c) IRGA (d) Atterberg's apparatus N

171. Which plant elements accounts for at least one half of the total number of ions absorbed-

(a) N (b) P

(c) K (d) Ca

172. The readily available form of K in soil ranges from—

(a) 0.1-2.0% (b) 1.0-10%

(c) 30-40% (d) 90-98%

173. CEC of mica is

(a) 0 (b) 30-50

(c) 80-100 (d) 150

174. S content (%) in potassium sulphate is -

(a) 50 (b) 22

(c) 17 (d) 12

175. Sulphur soil testing is determined by using -

(a) $CaCl_2$ (b) Hot water

(c) NH_4OAC (d) $NaHCO_3$

176. Grass tetany in forage crop is caused by deficiency of -

(a) Mg (b) S

(c) Ca (d) Mn

177. Which method is applied for testing of Fe, Zn, Cu and Mn -

(a) EDDHA (b) HEDTA

(c) Citric acid (d) DTPA

178. Which one of the following is not a Mn mineral -

(a) Pyrolusite (b) Hausmanite

(c) Rhodonite (d) Tenorite

179. Which one elements is the only non metal among the micronutrients-

(a) Mn (b) B

(c) Fe (d) Zn

180. Calcium hydroxide has a neutralizing value($CaCO_3$ equivalent) of -

(a) 179% (b) 136%

(c) 109% (d) 100%

181. Lime recommendation is made on the basis of a—— inch furrow slice -

(a) 2 (b) 4

(c) 6 (d) 15

182. Which of the following soils have poor physical condition -

(a) Saline (b) Sodic

(c) Saline –alkaline (d) None of these

183. Leaching requirement can be estimated by formula

(a) $LR=EC_{iw} / EC_{dw}$ (b) $LR=D_{iw} / D_{dw}$

(c) $LR=EC_{dw} / EC_{iw} \times 100$ (d) None of these

184. Diagnosis and recommendation integrated system (DRIS) is applied for

(a) Long duration crop (b) Short duration crop

(c) Oilseed crop (d) Vegetables

185. Who considered the father of Pedology
(a) CG Hopkins (b) JB Lawes
(c) VV Dokuchaev (d) Lie big

186. Who considered the various properties of soils in relation to plant production-
(a) Agronomist (b) Edaphologists
(c) Pedologist (d) Soil Chemists

187. Unconsolidated material overlying rocks is known as
(a) Solum (b) Regolith
(c) Bed rocks (d) A horizons

188. Vertical section of a soil is known as
(a) Horizons (b) Solum
(c) Profile (d) A soil

189. Least resistance mineral to weathering is
(a) Quartz (b) Olivine
(c) Albite (d) Gypsum

190. Which one of the following is a secondary mineral
(a) Muscovite (b) Anorthite
(c) Dolomite (d) Biotite

191. Inorganic parent material transported by gravity is
(a) Glacial (b) Eolian
(c) Marine (d) Colluvial

192. Passive factor of soil formation includes
(a) Climate (b) Organism
(c) Parent material (d) None of these

193. Lowest category of soil classification includes
(a) Order (b) Great group
(c) Series (d) Family

194. Which type of structure is dominant in grassland
(a) Spheroid (b) Platy
(c) Columnar (d) Blocky

195. Percent pore space can be calculated by using of which formula
(a) %PS= 100-(Dp / Db x 100) (b) %PS= 100-(Db / Dp x 100)
(c) %PS= 1-(Db / Dp x 100) (d) %PS= (Db / Dp x 100)

196. Which one of the following water potential in soil is always positive
(a) Osmotic potential (b) Metric potential
(c) Gravitational potential (d) None of these

197. In neutron moisture meter, source of fast neutrons is

(a) Radium
(b) Americium-beryllium
(c) Both (a) and (b)
(d) None of these

198. How much quantity of total incoming solar radiation is absorbed by the soil and can be used to warm

(a) 10% (b) 20%
(c) 42% (d) 50%

199. Magnesium in octahedral sheet of clay mineral indicates

(a) Diocthedral (b) Trioctahedral
(c) Tetrahedral (d) Monohedral

200. Aluminium in octahedral sheet of clay mineral indicates-

(a) Diocthedral (b) Trioctahedral
(c) Tetrahedral (d) Monohedral

201. Pressure membrane apparatus is used to measure metric potential moisture content relation at potential values as low as

(a) 1bar (b) 10 bar
(c) 100 bar (d) 1000 bar

202. Availability of which micro nutrient increases with increase in soil pH

(a) B (b) Fe
(c) Zn (d) Mo

203. 'White alkali' soils are

(a) Acid soil (b) Saline soil
(c) Saline sodic soil (d) Sodic soil

204. Which group of microorganism have highest biomass in soil

(a) Bacteria (b) Actinomycetes
(c) Fungi (d) Algae

205. Humus fraction lightest in colour is

(a) Fulvic acid (b) Humic acid
(c) Humin acid (d) None of these

206. Which group of colloidal have highest CEC by volume basis

(a) Humus (b) Vermiculite
(c) Smectite (d) Kaolinite

207. Zinc sulphate monohydrate contains Zn %

(a) 19 (b) 54
(c) 21 (d) 35

208. Boron content in borax (Sohaga) is
(a) 11% (b) 25%
(c) 19% (d) 39%

209. The smallest volume that can be called 'a soil' is
(a) Pedon (b) Ped
(c) Radon (d) Pans

210. Soils have no diagnostic pedogenic horizons are included in soil order
(a) Histosols (b) Mollisols
(c) Ultisols (d) Entisols

211. Flame photometer is used in the determination of
(a) Nitrogen (b) Phosphorus
(c) Potassium (d) Boron

212. The extractant used for available phosphorus in soil is
(a) Alkaline permanganate (b) 0.5M $NaHCO_3$
(c) NN Ammonium acetate (d) N Acetic acid

213. Soils having 200 kg available K_2O/ha will be classed as
(a) Low (b) Medium
(c) High (d) None of these

214. Sugar beet is an indicator plant for
(a) P (b) B
(c) Mg (d) Na

215. On the functional basis, Fe, Zn, Mn and Cu have been classified as
(a) Main structural elements (b) Accessory structural elements
(c) Regulators and carriers (d) Catalyst and activators

216. In Mitscherlich technique ——— plants grown to maturity.
(a) Barley (b) Oat
(c) Wheat (d) Sunflower

217. Calcium cyanamide contains nitrogen %
(a) 16 (b) 21
(c) 26 (d) 33

218. The term functional or metabolic nutrient was given by
(a) JV Lie big (b) DJ Nicholas
(c) DI Arnon (d) None of these

219. Rock formed by solidification of molten magma at the surface of the earth are called
(a) Sedimentary (b) Metamorphic
(c) Igneous (d) None of these

220. The rate of weathering is generally high in

(a) Arid tropic climate (b) Arid temperature climate
(c) Humid tropic climate (d) Humid temperate climate

221. The dominant spectral colour in soil indicates

(a) Hue (b) Value
(c) Chroma (d) Organic matter

222. The pH of the soil solution determines

(a) Active acidity (b) Exchange acidity
(c) Total acidity (d) None of these

223. The major source of negative charge on humus colloids is due to

(a) Carboxylic (-COOH) groups(b) Enolic (-OH) groups
(c) Phenolic groups (d) All of these

224. The value of ESP in alkali soils is always

(a) > 15 (b) <15
(c) Any value (d) None of these

225. Most dominant mineral in earth crust is

(a) Feldspar (b) Amphiboles
(c) Quartz (d) Mica

226. Munsell colour chart contains about ——different colored papers or chips

(a) 100 (b) 125
(c) 150 (d) 175

227. A mature and dry plant tissue contain about proteins (%)

(a) 20-25 (b) 1-15
(c) 1-8 (d) 10-30

228. The moisture equivalent was introduced by

(a) Brigg and Mc Lane (1907) (b) Richards and weaver(1944)
(c) Shaw and Baver(1939) (d) Whitney

229. Which type of soil survey use the lowest categories of taxonomic and mapping units (types and phases)

(a) Detailed survey
(b) Reconnaissance survey
(c) Detailed reconnaissance survey
(d) None of these

230. Flour apatite is a rich source of

(a) N (b) P
(c) K (d) Mo

231. Fe concentration used for foliar spray is

(a) 0.2% (b) 0.4%

(c) 0.05% (d) 1.0%

232. Equivalent acidity of anhydrous ammonia is

(a) 148 (b) 128

(c) 110 (d) 93

233. Molybdate reagent is used for estimation of

(a) N (b) P

(c) Ca (d) B

234. Who developed method for estimation of lime requirement-

(a) Sckooner (1952) (b) Shoemaker *et.al.*(1961)

(c) Stanford and English (1949)(d) Olsen *et.al* (1954)

235. Indicator plant of copper is

(a) Maize (b) Wheat

(c) Sorghum (d) Potato

236. *Aspergillus niger* is used to test the status of which elements in the soil

(a) N (b) P

(c) K (d) Fe

237. Which science assumes the status of petrology,Pedology and pedogenesis-

(a) Soil Science (b) Edaphology

(c) Pedology (d) Petrology

238. 2::1::1 type of clay mineral contains an extra sheet of —— as compared to 2::1 type of clay minerals

(a) Gibbsite (b) pyroxenes

(c) Brucite (d) Biotite

239. Which one of the following is a secondary mineral

(a) Muscovite (b) Anorthite

(c) Dolomite (d) Biotite

240. Chemical formula of siderite is

(a) $CaCO_3$ (b) Fe_2O_3

(c) $Ca_3(PO_4)_3$ (d) $FeCO_3$

241. Select the correct sequence of weathering stages in soil formation

(a) Juvenile-initial-virile-senile-final

(b) Senile-virile- Juvenile-initial-final

(c) Initial -Juvenile-virile-senile-final

(d) Initial- virile- Juvenile-senile-final

242. Silica:: sesquioxide (R_2O) ratio for tropical stages in soil formation-
(a) >3.5 (b) 3-3.5
(c) 2-3 (d) <2

243. Hydromorphic soils are evolved by which process
(a) Calcification (b) Gleization
(c) Salinization (d) Solonization

244. Solonization refers to the removal of ——— from the exchanges sites
(a) Ca^{++} (b) Mg^{++}
(c) Na^{+} (d) NaCl

245. Pedoturbation is the process of
(a) Mixing of the soil (b) Mixing of FYM in soil
(c) Mixing of fertilizer in the soil(d) Mixing of pesticides

246. Permafrost is a layer where soil temperature is always
(a) d″O ^{0}C (*b*) ‡ O^{0}C
(c) 0 to100 (d) d″10°C

247. Which soil order have highest area in India
(a) Entisols (b) Inceptisols
(c) Vertsols (d) Afisols

248. In India, which soil order have negligible area
(a) Oxisoils (b) Ultisols
(c) Mollisols (d) Histosols

249. Which mineral substances placed at number 10 in mh0's scale
(a) Talc (b) Gypsum
(c) Carborundum (d) Diamond

250. Which rock is volcanic (extrusive) in native
(a) Gabbro (b) Diorite
(c) Basalt (d) Monzonite

251. When organic matter is greater than 1.5% in soil rated as
(a) Low (b) medium
(c) High (d) Very high

252. A soil profile having two sola, one above the other is known as-
(a) Bisequem profile (b) Truncated profile
(c) Soil monolith (d) Soil fabric

253. Chernozem soils are found in
(a) India (b) USA
(c) Russia (d) Brazil

254. Which of the following mineral is non expending 2::1 type mineral

(a) Montmrillonite (b) Beidellite

(c) Nontronite (d) Illite

255. Humus forming fungi includes which genera-

(a) Trichoderma (b) Alternaria

(c) Aspergillus (d) Both (a) and (b)

256. Red soils are mainly confined to

(a) UP (b) Rajasthan

(c) Tamilnadu (d) Bihar

257. Thermal conductivity follows the order-

(a) Sand>Loam>Clay>Peat (b) Sand>Peat >Loam >Clay

(c) Loam>Sand >Peat>Clay (d) Peat> Sand>Loam>Clay

258. "Law of restitution was proposed

(a) Olsen (b) Das

(c) Lie big (d) Bray

259. Nutrient index between 1.5 and 2.5 is categorized as

(a) Low (b) Medium

(c) High (d) None of these

260. The coordinated project on "long term fertilizer experiments" (LTFE) was sponsored by ICAR in

(a) 1970-71 (b) 1975-76

(c) 1980-81 (d) 1985-86

261. Total number of master horizons present in soil are

(a) 4 (b) 5

(c) 6 (d) 7

262. The B horizons is generally absent in-

(a) Black soils (b) Red soils

(c) Desert soils (d) Brown hill soils

263. Biosuper fertilizer containing

(a) Rock phosphate + elemental sulphur+ S. Oxidizing bacteria

(b) Phosphoric acid + elemental sulphur+ S. Oxidizing bacteria

(c) Ammonium sulphate + elemental sulphur+ S. Oxidizing bacteria

(d) None of these

264. Fertilizer control order issued in

(a) 1947 (b) 1957

(c) 1977 (d) 1986

265. Gapon's equation is widely used equation to describe
(a) P status of soils (b) CEC of soils
(c) Organic matter of soils (d) Water status of soils

266. Who established the essentiality of Fe for plant in 1844
(a) Gilbert (b) Glauber
(c) Gris (d) Grigg

267. Majors deposits of gypsum are found in
(a) Rajasthan (b) Haryana
(c) Gujarat (d) Punjab

268. Which formula is applied to calculate gypsum requirement (meq/ 100 gm)
(a) $GR= ESP_{final} - ESP_{initial}/100 \times CEC$
(b) $GR= ESP_{initia} - ESP_{final}$ 1/ $CEC \times 100$
(c) $GR= ESP_{final} - ESP_{initial}/ CEC \times 100$
(d) $GR= ESP_{final} - ESP_{initial}/ AEC \times 100$

269. Who developed the DTPA tests for available soil micronutrients
(a) Kurtz LT
(b) Lindsay WL
(c) Larsen
(d) Jackson ML

270. The value of Van Bemmelen factor is
(a) 0.43 (b) 1.200
(c) 2.290 (d) 1.724

271. One hectare cm of water is equal to
(a) 100 tones of water (b) 100000 tones of water
(c) Both a and b (d) 10 tones of water

272. Journal "Plant and Soil" is published from
(a) New Delhi (b) The Netherlands
(c) Japan (d) USA

273. The relative proportion of sand, silt and clay in a soil is called
(a) Soil texture (b) Soil structure
(c) Soil taxonomy (d) Soil aggregation

274. 1 bar is equal to
(a) 0.9869 atm
(b) Weight of a 1020 cm water column
(c) 75.01 cm high mercury column
(d) All of these

275. The tension of the water at the upper plastic limit is equivalent to pF value

(a) 0.5 (b) 1.0
(c) 2.0 (d) 2.8-3.3

276. Source of negative charges on silicate clays are

(a) Exposed crystal edges (b) Isomorphous substitution
(c) Anion exchange (d) All of these

277. Browning disease of rice is a

(a) Physiological disease (b) Fungal disease
(c) Bacterial disease (d) Viral disease

278. Walkley and Black's method of organic matter determination allows———% recovery

(a) 54 (b) 64
(c) 74 (d) 84

279. Water stable aggregates in soil increase by addition of

(a) N-fertilizers (b) P-fertilizers
(c) K-fertilizers (d) Micronutrients

280. A soil which has pH <8.5, ESP <15 and EC>4 dSm^{-1} at 25 °C is called

(a) Saline soil (b) Alkaline soil
(c) Saline-Alkaline soil (d) None of these

281. Fertility gradient approach was given by

(a) B V Mehta (b) B. Ramamurthy
(c) AB Ghosh (d) B Murthy

282. Hydrogen bonding takes place in the interlayer of

(a) Kaolinite (b) Montmrillonite
(c) Mica (d) Chlorite

283. Ionic potential is maximum in case of

(a) Na (b) Cs
(c) K (d) Li

284. Which one of the following is commonly used as urease inhibitor

(a) Dicyandiamide (b) Thiourea
(c) Urea formaldehyde (d) Phenol phosphorodiamidate

285. India agriculture is operating at a negative balance of plant nutrients

(a) 2-5 mt (b) 8-10 mt
(c) 10-15 mt (d) 20-25 mt

286. Which country has largest deposit of K

(a) India (b) Canada
(c) USA (d) China

287. Which country has monopoly in P deposits

(a) Morocco (b) South Africa
(c) Russia (d) Pakistan

288. In India, which fertilizer manufactured first

(a) Urea (b) DAP
(c) SSP (d) ASN

289. Hartig nets in soil produce by

(a) Rhizobium (b) Ectomycorrhiza
(c) Endomycorrhiza (d) None of these

290. End product of denitrification process is

(a) N_2 (b) NO_3
(c) N_2O and N_2 (d) NH_3

291. Which nitrogenous fertilizer industrially made first

(a) $CaCN_2$ (b) Ammonium sulphate
(c) $NaNO_3$ (d) Urea

292. The losses of ammonium usually become greater with

(a) Decrease in calcium (b) Decrease in temperature
(c) Decrease in soil pH (d) Increase in soil pH

293. The highest percentage content of nitrogen among the commonly used nitrogenous fertilizers is embodied by

(a) Ammonium Nitrate (b) Urea sulphate
(c) Urea (d) Anhydrous ammonia

294. Phosphotase enzymes play a major role in ——organic phosphates in soil

(a) Immobilization (b) Mass transportation
(c) Mineralization (d) None of these

295. A minor portion of the phosphorus sorption capacity of soils originated from

(a) Sodium bicarbonate (b) Calcium carbonate
(c) Calcium oxide (d) Sodium carbonate

296. The maximum phosphorus availability in most of the soils is in the pH range

(a) 4.0-4.5 (b) 7.0-7.5
(c) 6.0-6.5 (d) 5.0-5.5

297. ——————generally augments the utilization of phosphorus which is water soluble-

(a) Broadcast (b) Broadcast and incorporation
(c) Band placement (d) None of these

298. The essentiality of copper for higher plants was first recognized in ———.

(a) 1921 (b) 1931
(c) 1941 (d) 1911

299. "Reclamation disease" due to copper deficiency occurs in plants grown on newly reclaimed acid

(a) Histosols (b) Alfisols
(c) Ultisols (d) Aridisols

300. Repeated use of Bordeaux mixture, sewage sludge pig poultry manures and /or mine waste may cause-

(a) Fe (b) Al
(c) Cu (d) Zn

301. The ratio of organic carbon to organic matter in soil is

(a) 1.0-2.0 (b) 20.0-1.0
(c) 1.0-1.7 (d) 20.7-1.0-

302. The bulk density of a soil with particle density of 2.65g/cc and pore space 40% would be- (g/cc)

(a) 1.29 (b) 1.59
(c) 1.45 (d) 1.72

303. Iron deficiency usually occurs in soil in the range of

(a) 5.4-6.5 (b) 7.4-8.5
(c) 9.6-10.1 (d) 9.0-9.5

304. Silicon is the ——most abundant element in the earth's crust

(a) Third (b) Sixth
(c) Second (d) Fourth

305. Nitrogen fertilizer use efficiency in rice can be increased by using-

(a) Sulphur coated urea (b) Sulphur-coated urea
(c) BGA (d) Urea super granules

306. Which one of the following nutrients plays the most effective role in controlling the rate of transpiration

(a) P (b) Zn
(c) K (d) S

307. Plants roots are

(a) Positively phototropic (b) Positively geotropic
(c) Negatively hydrotropic (d) Negatively hydrotropic

308. When the phosphorus is either band placed or broadcast, the relative efficiency is related to —

(a) Its soil status (b) Its applied quantum
(c) Both (a) and (b) (d) None of these

309. Potassium salts are

(a) Less mobile than P (b) Less mobile than nitrate

(c) More mobile than nitrate (d) As mobile as nitrates

310. Broadcast potassium is —— than banded-

(a) More efficient (b) Less efficient

(c) As efficient as (d) None of theses

311. Winter application of phosphorus is followed in

(a) Warm areas (b) Cool areas

(c) Both (a) and (b) (d) None of these

312. Potassium is utilized more efficiently if applied

(a) Before/at planting (b) Side dressing

(c) Top dressing (d) None of these

313. Availability of P to crop plants is most effected by

(a) Soil pH (b) Temperature

(c) Biological factor (d) Soil moisture stress

314. Nitrogen in DAP

(a) 36% (b) 9%

(c) 18% (d) 46%

315. In mustard crop, gypsum is applied to supply

(a) Sulphur (b) Calcium

(c) Nitrogen (d) Zinc

316. Among oil cakes the highest nitrogen content is in

(a) Neem cake (b) Castor cake

(c) Coconut cake (d) Groundnut cake

317. Availability of Fe to plants is relatively high at the pH range

(a) 6.0-6.5 (b) 7.6-8.0

(c) 6.6-7.0 (d) 7.1-7.5

318. Which fertilizer is suitable and economical to apply under acidic condition.-

(a) Single super phosphate (SSP)(b) DAP

(c) Rock phosphate (d) Basic slag

319. In saline soil major problem is due to

(a) High exchangeable sodium (b) Low soil pH

(c) High pH (d) High content of soluble salts

320. Opening of stomata is related with

(a) K (b) Fe

(c) Ca (d) Mg

321. Symptoms of Fe deficiency in plants are
(a) Yellowing of whole plants (b) Yellowing of lower leaves
(c) Drying of lower leaves (d) Yellowing of young leaves

322. Which of the absorbed cations are largely responsible for soil acidity
(a) Hydrogen (b) Sodium
(c) calcium (d) Aluminium

323. Which of the soils have electrical conductivity (EC) less than 4dS/m.
(a) Acid (b) Saline
(c) Saline-sodic (d) Sodic

324. Which of the crops are salt sensitive crop
(a) Barley (b) Onion
(c) Potato (d) Sugar beet

325. Which of the following are metamorphic rocks
(a) Schist (b) Slate
(c) Dolomite (d) Shales

326. Addition of organic matter to soil
(a) Increase lower plastic limits(b) Increase upper plastic limits
(c) Increase plasticity index (d) Does not affect plasticity index

327. The earth worn casts care definitely higher in ——than the soil
(a) Bacteria (b) Organic matter
(c) Available nutrients (d) All of these

328. Use of phospho-gypsum has been found to be useful in the amelioration of
(a) Sub soil acidity (b) Alkali soils
(c) Saline soils (d) Acid sulphate soils

329. Arrange the Mn source in increasing order of Mn % age.
(a) Mn-EDTA (b) Manganese oxide
(c) Manganese chloride (d) Manganese sulphate

330. Arrange sulphur fertilizers in increasing percentage of sulphur content
(a) Ferrous sulphate (b) Ammonium sulphate
(c) SSP (d) ASN

331. Arrange in decreasing order of CEC
(a) Illite (b) Humus
(c) Vermiculite (d) Kaolinite

332. Silica-sesquioxide ratio is useful in the evaluation of soil in respect of
(a) Silica content (b) Nutrient availability
(c) Its age (d) Fe and Al content

333. Non-essential but useful nutrients for plants would include

(a) V and Na (b) V and Cl
(c) C and V (d) Na and Cl

334. If the land utilization index (LUI) of a crop is 0.50 than the duration of that crop will be

(a) 183 days (b) 135 days
(c) 200 days (d) 172 days

335. Which of the following elements is part of the chlorophyll?

(a) Fe (b) Mg
(c) Zn (d) Mn

336. Acid soils can be reclaimed by addition of

(a) Gypsum (b) Green manure
(c) Compost (d) Lime

337. Which elements is essential for nitrogen metabolism in green plants

(a) P (b) Zn
(c) Mo (d) K

338. Which elements is immobile in the plants

(a) K (b) N
(c) Ca (d) P

339. Terrace cultivation is mostly practiced

(a) In rural areas (b) In desert land
(c) On slop of hill (d) On slop of mountain

340. The chemical characteristics feature of non-saline alkali soil is EC (d S/m) ESP pH

(a) <4 >15 >8.5 (b) <4 <15 >8.5
(c) >4 >15 >8.5 (d) >4 <15 <8.5

341. What is the upper crops tolerable limit of biuret in urea?

(a) 1.5% (b) 0.5%
(c) 1.0% (d) 3.0%

342. In India, which fertilizer manufactured first

(a) Urea (b) DAP
(c) SSP (d) ASN

343. The crop plants transpire more when there is

(a) High temperature and low humidity
(b) High humidity and high wind velocity
(c) High temperature and high humidity
(d) Low temperature and low humidity

344. Current all India NPK use ratio as about

(a) 5-3-1 (b) 15-5-2

(c) 10-3-1 (d) 3-2-1

345. An organism which converts nitrite to nitrate is

(a) *Azotobacter* (b) *Nitrobacter*

(c) *Nitrosomonas* (d) *Nitrogenase*

346. The efficiency of field irrigation would be more

(a) Sandy loam (b) loam

(c) Clay loam (d) Heavy loam

347. Pedology considers soil as a

(a) Natural body

(b) Medium for plant growth

(c) Powered rock mass of the earth's crust

(d) Purely a synthesized one

348. Water and mineral salts are Tran located in plants in the

(a) Xylem (b) Phloem

(c) Collenchyma (d) Parenchyma

349. The layers in soil are referred to as

(a) Hard pans (b) Horizons

(c) Profile (d) Pedocals

350. In soil science micelle stands for

(a) Natively charged silt particles with high CEC

(b) Colloidal particles carrying positive charges

(c) Microspores in-between textural particles

(d) Colloidal particle carrying negative charges_

351. Total porosity of a soil with 1.325 g/cc bulk density and 2.65 g/cc particle density is

(a) 40% (b) 60%

(c) 30% (d) 505

352. Which state is devoid of arid zone

(a) MP (b) AP

(c) Punjab (d) Maharastra

353. India Institute of soil science is located at

(a) Ludhiana (b) Hisar

(c) New Delhi (d) Bhopal

354. The diameter of very fine sand particles is

(a) Less than0.002 mm (b) 0.05-0.10 mm

(c) Less than 0. 02 mm (d) 0.01-0.05 mm

355. What culture should be given priority on groundnut cultivation
(a) Azospirilla (b) Rhizobia
(c) Mycorriza (d) Frankia

356. Alkali soils are treated with gypsum to correct the deficiency of
(a) S (b) Ca
(c) Zn (d) Mg

357. Which micronutrients is most deficient in Indo-Genetic alluvium soils
(a) Cu (b) Zn
(c) Fe (d) Mn

358. Which fertilizer is most beneficial for alkali soil-
(a) Urea (b) Ammonium nitrate
(c) Ammonium sulphate (d) CAN

359. The most productive normal soil have pH
(a) 5-7 (b) 7-9
(c) 4-6 (d) 6-8

360. Which is the most harmful salt in salt affected soil
(a) Sodium carbonate (b) Sodium sulphate
(c) Sodium chloride (d) Sodium bicarbonate

361. Which elements is not phototoxic
(a) Se (b) Li
(c) Fl (d) B

362. Which fraction of humus is soluble both in alkali and alcohol
(a) Fluoric acid (b) Humic acid
(c) Hymatomelanic acid (d) Humin

363. Which of the following is not the advantage of using mixed fertilizers
(a) Economical (b) Require less care
(c) Application is easier (d) Reduced salinity

364. Which crop has maximum nitrogen fixing potential?
(a) Pigeon pea (b) Soybean
(c) Guar (d) Groundnut

365. Maximum area of black soils in India is in
(a) Maharastra (b) MP
(c) Rajasthan (d) UP

366. Which fruits tree is sensitive to acidity and salinity
(a) Plum (b) Guava
(c) Phalsa (d) Aonla

367. In India, which fertilizer manufactured first

(a) Urea (b) DAP
(c) SSP (d) ASN

368. The process of soil formation include—

(a) The addition of organic and mineral matter
(b) The translocation of materials from one point to another within the soil callus
(c) The loss of OM and mineral material from the soil
(d) All of the above

369. If the nitrogen content of a soil is 0.56%, the organic matter (%) content of that soil will be-

(a) 5.6 (b) 0.95
(c) 9.5 (d) 1.72

370. The nitrogenous fertilizer manufactured in India which occupies the first place in production to-day is

(a) Urea (b) Ammonium sulphate
(c) Ammonium phosphate (d) CAN

371. Which fertilizers is least hygroscopic

(a) Ammonium sulphate (b) Calcium ammonium sulphate
(c) Ammonium phosphate (d) Urea

372. DAP requirement for supplying 45 kg N and 115 kg P_2O_5 to one hectare peanut crop is

(a) 150 kg (b) 250 kg
(c) 200 kg (d) 100 kg

373. Chlorosis in plants is caused due to deficiency of

(a) Mg and S (b) B and P
(c) N and Mg (d) Ca and O_2

374. Fruit formation is suppressed due to deficiency of

(a) B (b) Mg
(c) S (d) Ca

375. In photosynthesis, oxygen is liberated due to

(a) Breakdown of CO_2
(b) Reduction of CO_2
(c) Oxidation of H_2O
(d) Breakdown of chlorophyll

376. The plants which manufacture their own food in presence of

(a) Photoautotrophic (b) Phototropic
(c) Autotrophic (d) Photophytic

377. In India, which fertilizer manufactured first

(a) Urea (b) DAP

(c) SSP (d) ASN

378. Photosynthesis does not take place in fungi due to absence of

(a) Chlorophyll (b) K

(c) N (d) P

379. The strongest chemical agent of breaking dormancy is

(a) Gibberellin (b) Cumerin

(c) Ethylene (d) KNO_3

380. The optimum cardinal temperature points for germination of rice seeds are

(a) 20-22°C (b) 18-20°C

(c) 30-32°C (d) 37-39°

381. Interveinal chlorosis of the younger leaves of the plants is the characteristics of deficiency symptoms of

(a) P (b) Zn

(c) Fe (d) B

382. The waterlogged saline soil can be reclaimed through

(a) Chemical amendment (b) Fertilizers

(c) Biodrainage (d) Farmyard manure

383. Total dissolve solid in water having EC of 5 dS/m will be about (Hint:: 64x 10^3 x 5 = = 3200 ppm)

(a) 3200 ppm (b) 4400 ppm

(c) 4000 ppm (d) 3600 ppm

384. The Sodium Absorption Ratio (SAR) of the soil containing Na, Mg and Ca ions as 100,30 and 20 meq/L respectively, will be about

(a) 10 (b) 15

(c) 15 (d) 5

385. Sodic soils generally exhibit soil structure of the type

(a) Platy (b) Granulated

(c) Sub-angular blocky (d) Prismatic /Columnar

386. The residual sodium carbonate of water containing carbonate, bicarbonate Ca and Mg, ions as 5,35,20,10 and meq/L respectively will be (in meq/L)

(a) 0 (b) 5

(c) 10 (d) 20

387. Soil water content(%) at field capacity is maximum / highest with

(a) Loam (b) Silt loam

(c) Clayey soils (d) Clay loam

388. Carryover fertilizer effects are more pronounced in the case of

(a) Urea (b) CAN

(c) Ammonium nitrate (d) SSP

389. Laterite soils occurs in the district of

(a) Indore district of MP

(b) Bikaner district of Rajasthan

(c) Patiala district of Maharashtra

(d) Ratnagiri district of Maharashtra

390. Journal of Soil Science is published from

(a) India (b) Canada

(c) Australia (d) USA

391. X-ray diffraction technique is used to study

(a) Tracer study

(b) Photosynthesis

(c) Clay mineralogy

(d) None of these

392. Biuret content in urea fertilizer is permitted up to (ECO)

(a) < 2.5% (b) Between 2-3%

(c) Maximum up to 1.0% (d) 0

393. Alkali soil can be reclaimed by the addition of

(a) Pyrites (b) Basic slag

(c) Lime (d) Calcite

394. At 1 x 10^{-9} H^+ concentration, what will be the pH value

(a) 6 (b) 5

(c) 10 (d) 9

395. Anion exchange capacity is found minimum in the case of

(a) Montmorillonite (b) Kaolinite

(c) Illite (d) None of these

396. Cation exchange capacity is found minimum in the case of

(a) Montmorillonite (b) Humus

(c) Illite (d) Kaolinite

397. Chemical weathering governs mostly by the process of

(a) Hydration (b) Hydrolysis

(c) Solution (d) All of these

398. Limonite is formed by the process of

(a) Oxidation (b) Hydrolysis

(c) Carbonation (d) None of these

399. Ferrous oxide can be obtained during chemical process weathering under the head of

(a) Solution (b) Hydration

(c) Oxidation (d) None of these

400. The systemic examination, description, classification and mapping of soil of an area is known as

(a) Soil survey (b) Soil classification

(c) Soil mapping (d) All of these

401. All India soil survey and Land Use Organization was established in

(a) 1962 (b) 1952

(c) 1956 (d) 1960

402. Which acts also as nitrification inhibitor over and above a manure

(a) Ground nut (b) Sesamum

(c) Neem (d) Karange

403. The mineral chlorite has an extra layer of

(a) Brucite (b) Gibbsite

(c) Montmorillonite (d) Illite

404. Which one of the following minerals of the main source of boron in soil-

(a) Tourmaline (b) Hemimorphite

(c) Hematite (d) Pyrolusite

405. Criteria of essentiality of nutrients for plants were given by

(a) Arnon (b) Russed

(c) Stout (d) Lie big

406. Appetite is a group of mineral which mainly contains

(a) S (b) Ti

(c) N (d) P

407. The diagnostic horizon in Alfisols termed as

(a) Spodic (b) Oxic

(c) Argillic (d) Mollic

408. Which soils has got the highest bulk density-

(a) Clay loam (b) Silty loam

(c) Loam (d) Sandy soil

409. The equation of soil formation, S= f (Cl, o, r, p, t) was given by

(a) Glinka (b) Jenny

(c) Kelong (d) Dokuchaev

410. A soil formation process in which humus and sesquioxides become mobile and remain behind

(a) Calcification (b) Podzolization

(c) Pedoturbation (d) Humification

411. Attraction of solid surface for water molecule is called

(a) Adsorption (b) Adhesion

(c) Cohesion (d) Absorption

412. Erosion caused by spattering of soil particles by the impact of rain drops is

(a) Sheet erosion (b) Gully erosion

(c) Rill erosion (d) Sparsh erosion

413. Which is not a sedimentary rock

(a) Lime stone (b) Sandstone

(c) Conglomerate (d) Gneiss

414. Chlorite belongs to the

(a) 2::2 Type (b) 1::1 type

(c) 2::1 type (d) Amorphous type

415. The soil layer of maximum leaching is indicated by the symbol-

(a) A_1 (b) A_2

(c) B_1 (d) B_2

416. National Remote Sensing Agency(NRSA) was established in

(a) New Delhi,1977 (b) Hyderabad,1968

(c) Bombay,1973 (d) Banglore,1970

417. The diameter of the silt particles is

(a) 0.0005-0.0002 mm (b) 0.5-0.2 mm

(c) 0.002 mm (d) 0.05-0.002

418. Different soil separate may be determined with the help of

(a) Yet's chart (b) Munsell chart

(c) International pipette method(d) None of these

419. The natural soil aggregates are called as

(a) Soil separates (b) Mottling

(c) Pads (d) Concretion

420. The average soil bulk density of cultivated soil is approximately in (g/cm^3)

(a) 1.5-1.65 (b) 1.1-1.4

(c) 0.5-1.0 (d) None of these

421. A combination of many taxonomic unit like soil series, phase or type is called

(a) Soil series
(b) Soil complexes
(c) Soil vibrant
(d) Soil association

422. The patterns of occurrence of various soil taxonomic units on the details map is called

(a) Soil phase (b) Soil association
(c) Soil vibrant (d) Soil complexes

423. The soil texture of surface soil of surface horizons is called

(a) Soil phase (b) Soil variant
(c) Soil type (d) ASN

424. In India, which fertilizer manufactured first

(a) Urea (b) DAP
(c) None of these (d) ASN

425. The representation of surface feature on map and their identification is called

(a) Soil variant (b) Soil map
(c) Soil legend (d) None of these

426. An abrupt change in texture class of no continuity found between B and C horizons is termed as

(a) Litho logical discontinuity (b) Regolith discontinuity
(c) Textural change (d) None of these

427. How many times the soil organic matter hold the water of it is weight

(a) 20 Times (b) 15 Times
(c) 10 times (d) 25 times

428. Water drains by gravitational force from pores larger than

(a) About 30-60 microns (b) About 20-30 microns
(c) About 10-20 micron (d) None of these

429. Soils with high pH are generally deficient in

(a) Zn and Mn (b) Ca and Mg
(c) B and Fe (d) Cu and Mo

430. Which one plants nutrients is useful in increasing resistance to disease and insect pest

(a) Ca (b) P
(c) K (d) N

431. Essential elements are those
(a) Which could be replaced by other elements with similar properties
(b) Which are required in large quantities
(c) Whose deficiency would prevent the completion of plants life cycle
(d) Which would promote plant growth

432. Molybdenum is required by plants because it is a co-factor for
(a) Cytochrome oxidase (b) Phosphotase
(c) Nitrite reductase (d) Nitrate reductase

433. Khaira disease of rice can be controlled by spraying of
(a) Copper sulphate (b) Manganese sulphate
(c) Zinc sulphate (d) Borax

434. Organic matter content of Histosols and Hydromorphic soil is
(a) 30-40% (b) 2-18%
(c) 50-57% (d) 60-70%

435. The relative amounts of air and water in the pore spaces:-
(a) Half filled and half empty (b) Always remain constant
(c) Fluctuate continuously (d) All of these

436. Moisture content from lower to upper plastic limit is known as-
(a) Sticky point (b) Plastic number
(c) Field capacity (d) All the above

437. The main source of heat energy for soil is
(a) Infra red radiation (b) Solar radiation
(c) Ultra violet radiation (d) None of these

438. Specific heat of the mineral matter is (cal / g)
(a) 1.0 (b) 24.0
(c) 0.462 (d) 0.18

439. Example of cumulous material are
(a) Gneiss (b) Till plain
(c) Peat and Muck (d) Colluvial

440. Aeolian deposits are formed by the transportation of material through
(a) Wind (b) Sea
(c) Running water (d) Ice

441. A vertical section of soil body which show different layer is called
(a) Soil profile (b) Regolith
(c) Soil horizon (d) Parent material

442. Solum is made up of

(a) C horizon (b) C+D horizon
(c) A+B horizon (d) E horizon

443. Identify which is a passive soil forming factor

(a) Relief (b) Climate
(c) Biosphere (d) None of these

444. In India, which fertilizer manufactured first

(a) SSP (b) DAP
(c) SSP (d) ASN

445. Aridisols soil are found in the region of

(a) Humid (b) Submerged
(c) Arid (d) None of these

446. The dominant colour of spectral (Rain bow) in soil indicates

(a) Clay mineral (b) Value
(c) Hue (d) Chroma

447. Montmorillonite is a type of clay mineral

(a) 2::2 (b) 3::1
(c) 1.1 (d) None of these

448. Soils are acidic for one of the reason

(a) Acidic parent material (b) Basic slag
(c) Low rain fall (d) None of these

449. A soil having pH 8.5;ESP 15 and EC <4 dS/m at 25 ^oC is known

(a) Saline soil (b) Alkali soil
(c) Recent alluvium (d) Saline –alkali soil

450. A soil having pH 6.0 is called

(a) Acidic soil (b) Saline soil
(c) Alkaline soil (d) Normal soil

451. The diameter of water molecule is approximately

(a) 2.64^{o}A (b) 2.54^{o}A
(c) 1.33^{o}A (d) 2.94^{o}A

452. How many order are included in the 7th approximation system of soil classification

(a) 8 (b) 10
(c) 12 (d) 15

453. Clays are classified as

(a) Primary minerals (b) Hydroxy minerals
(c) Secondary minerals (d) Tertiary minerals

454. Soil formed under impeded drainage condition where movement of air and water is restricted is known

(a) Acid soil (b) Podzol

(c) Gley soil (d) Chernozem

455. Physical weathering causes

(a) Alternation of rocks and minerals

(b) Decomposition of rocks and minerals

(c) Disintegration of rocks and minerals

(d) Synthesis of rocks and minerals

456. If the C::N ratio of crop residues is 20-30 then

(a) Partial mineralization may not take place

(b) Immobilization may not take place

(c) Mineralization may take place

(d) Not known

457. If the C::N ratio of crop residues is more than 30 than

(a) Partial mineralization may take place

(b) Immobilization will take place

(c) Mineralization may take place

(d) All of these

458. If the C::N ratio of crop residues is less than 22 than

(a) Mineralization will take place

(b) Immobilization will take place

(c) Both of these

(d) None of these

459. The organism which can survive at the temperature 10-15°C is called

(a) Psychrophylic (b) Mesophylic

(c) Thermophylic (d) Thermoduric

460. The microbes which can survive at the temperature 25-30°C is called

(a) Thermoduric (b) Thermophylic

(c) Mesophylic (d) Psychrophylic

461. Which nitrogenous fertilizer will you recommend for paddy crop

(a) Nitrate fertilizer (b) Nitrate ammonium fertilizer

(c) Amide fertilizer (d) All of the above

462. In India, which fertilizer manufactured first

(a) Urea (b) DAP

(c) SSP (d) AS

463. The common chemical preservatives are

(a) Calcium chloride (b) Calcium sulphate

(c) Magnesium sulphate (d) None of these

464. The residual effect of ammonical nitrogenous fertilizer in soil as

(a) Alkaline (b) Acidic

(c) Neutral (d) None of these

465. Urea is mineralized in the presence of enzyme

(a) Peptase (b) Urease

(c) Co-enzyme (d) All of the above

466. When you apply 80 kg N/ha to wheat crop how many kg of urea would be required

[Hint:: 2.17 x 80= 173.9 kg]

(a) 120.0 kg (b) 80.00 kg

(c) 173.93 kg (d) 100.25 kg

467. Single super phosphate (SSP) is manufactured from

(a) Rock phosphate (b) Bone meal

(c) Di- calcium phosphate (d) ASN

468. Rhizobium species and the legume is called

(a) Non-symbiotic N_2-fixation (b) Symbiotic N_2-fixation

(c) Both (a) and (b) (d) None of these

469. *Rhizibium japonicum* fix the nitrogen in the root of

(a) Soybean (b) Pea

(c) Beans (d) Lentil

470. The N_2 fixation in the roots of Alfalfa is———kg N/ha/yr.

(a) 100-150 (b) 50-75

(c) 125-300 (d) 150-200

471. The nitrogen fixation in soybean may be up to ——— kg N/ha/ year

(a) 65-115 kg (b) 60-100 kg

(c) 100-150 kg (d) None of these

472. The N_2 fixation by *Azotobacter* and *Azospirillum* in the cereals is called

(a) Non symbiotic N_2 fixation (b) Symbiotic N_2 fixation

(c) Both of these (d) None of these

473. Percentage of soil solids in soil is

(a) 50% (b) 40%

(c) 30% (d) 60%

474. Which of the following is known as father of Pedology

(a) VV Dokuchaev (b) DG Vilensky

(c) KD Glinka (d) CF Marbut

475. Soil profile may be defined as

(a) Horizontal section of soil (b) Upper most layer of earth's crust
(c) The land space (d) Vertical section of soil

476. The illuviation is zone of

(a) Hard crust (b) Leaching
(c) Accumulation (d) None of these

477. The maximum moisture is available to plant at

(a) Wilting point (b) Hygroscopic coefficient
(c) Saturation (d) Field capacity

478. Soil air contain

(a) 2.5% CO_2 (b) 0.02% CO_2
(c) 3.0% CO_2 (d) 0.25% CO_2

479. The diagnostic surface horizon are called

(a) Polypedon (b) Pedon
(c) Epipedon (d) All of them

480. A thick dark coloured horizon which have high base saturation (50% or more) high organic matter content and with strong soil structure is called

(a) Ochric Epipedon (b) Histic Epipedon
(c) Umbric Epipedon (d) Mollic Epipedon

481. An epipedon which has dark colour low in base saturation is called

(a) Ochric (b) Histic
(c) Umbric (d) Mollic

482. A man made epipedon which is similar to mollic epipedon in many characters expect P_2O_5 content(more 250 ppm) is

(a) Umbric (b) Anthropic
(c) Ochric (d) Histic

483. In India, which fertilizer manufactured first

(a) Urea (b) DAP
(c) SSP (d) ASN

484. Geological weathering takes place in

(a) Lower layer of soil (b) On earth crust
(c) None of these

485. Bone meal is well suited for

(a) Alkaline soil (b) Neutral soil
(c) Acidic soil (d) All of these

486. Muriate of potash contain (% K_2O)

(a) 40 (b) 50
(c) 58 (d) None of these

487. Whiptail symptom in cauliflower is due to the deficiency of

(a) Fe (b) Mo
(c) Zn (d) Cu

488. The number of essential elements required for the growth of most higher plant is

(a) 8 (b) 17
(c) 16 (d) 3

489. For preparing a mixed fertilizer, 100 kg of ammonium sulphate and 125 kg of SSP mixed and 25 kg of filler added to this mixture. In terms of N-P_2O_5-K_2O the fertilizer grade of this mixture will be

(a) 10-10-0 (b) 5-10-0
(c) 8-8-0 (d) 5-5-0

490. A soil with 20 meq/100 g soil and 5meq/100 g exchangeable soil H will have a base saturation of

(a) 25.0% (b) 133.3%
(c) 80% (d) 75.0%

491. High RSC of irrigation water results

(a) Precipitation of Na (b) Precipitation of K
(c) Precipitation of Ca and Mg (d) None of these

492. Olsen's method of P determination is more effective in

(a) Normal soil (b) Alkaline soil
(c) Acidic soil (d) None of these

493. Mass flow of gases into the soil occurs due to the difference of

(a) Surface tension (b) Pressure gradient
(c) Both of above (d) None of these

494. Upper plastic limit is also known as

(a) Lower limit (b) Liquid limit
(c) Index of plasticity (d) None of these

495. If a soil has a bulk density of 1.50 mg/m^3 and a particle density of 2.65 Mg/m^3 then the percentage pore space of that soil will be

(a) 43.4 (b) 38.5
(c) 48.5 (d) 35.5

496. The C::N ratio of FYM is

(a) 20::1 (b) 10::1
(c) 30::1 (d) 40::1

497. Soil test should be repeated after every

(a) Year (b) 4 year
(c) 10 year (d) Once in enough

498. Plant analysis is useful if done at

(a) Harvest (b) Grain formation

(c) Early stage (d) Both (a) and (b)

499. The N fertilizer in all fertilizers come from

(a) Naphtha (b) Gas

(c) Air (d) Water

500. The most common phosphate fertilizer in India is

(a) SSP (b) 12-32-16

(c) DAP (d) Rock-P

501. Most important source of sulphur for India soil is

(a) Gypsum (b) AS

(c) Pyrite (d) SSP

502. SSP can have an edge over DAP if the soil is

(a) Deficient in P (b) Calcareous

(c) Water logged (d) Deficient in S

503. SOP is batter than MOP when the crops are

(a) Tall (b) N-deficient

(c) Sensitive to chloride (d) Legumes

504. CAN is not acid forming fertilizer because it contains-

(a) Ammonium (b) Nitrate

(c) Calcium (d) Granules

505. Nitrogen is applied in splits to

(a) Increase its efficiency (b) Help the dealer

(c) Reduce the credit load (d) All of above

506. Efficiency of urea can be increased by

(a) Deep ploughing (b) Using Neem cake

(c) Early harvesting (d) Both (a) and (b)

507. If out of 120-60-60, 80 kg N is to be top dressed the basal dose is

(a) 40-60-60 (b) 120-60-60

(c) 200-60-60 (d) 100-60-600

508. Fertilizer use is profitable if each rupee spent on fertilizer produces crop worth at least:

(a) Rs.1 (b) Rs.10

(c) Rs.5 (d) Rs.2

509. If fertilizer prices go up the optimum dose should

(a) Come down (b) Stay unchanged

(c) Also go up (d) None of these

510. India rank in fertilizer consumption in the world is

(a) 6 (b) 10

(c) 21 (d) 3

511. The state using largest tonnage of fertilizer is

(a) Punjab (b) AP

(c) UP (d) TN

512. The state which uses almost as much potash as N is

(a) Bihar (b) Gujarat

(c) Kerala (d) TN

513. The least hygroscopic fertilizes is

(a) SOP (b) CAN

(c) SSP (d) NPK complex

514. The Central Fertilizer Quality Control Institute is located at

(a) Delhi (b) Hyderabad

(c) Pune (d) Faridabad

515. Too much micro-nutrient application can produce

(a) Tall plants (b) Root death

(c) Toxicity (d) Big flower

516. Cereal straw is a particularly rich source of

(a) Na (b) K

(c) Zn (d) Fe

517. Fertilizer needed to make 300 litre of 1% solution is

(a) 3 kg (b) 1 kg

(c) 30 kg (d) 300 kg

518. Green manure are mostly used as a source of

(a) Zn (b) Carbon

(c) Chlorophyll (d) N

Answers

1. (d)	2. (a)	3. (a)	4. (a)	5. (c)
6. (b)	7. (c)	8. (a)	9. (d)	10. (d)
11. (c)	12. (a)	13. (a)	14. (b)	15. (c)
16. (a)	17. (b)	18. (a)	19. (b)	20. (c)
21. (c)	22. (c)	23. (c)	24. (c)	25. (d)
26. (b)	27. (d)	28. (a)	29. (b)	30. (b)
31. (b)	32. (c)	33. (d)	34. (a)	35. (b)
36. (b)	37. (b)	38. (c)	39. (a)	40. (a)
41. (c)	42. (a)	43. (d)	44. (a)	45. (c)
46. (c)	47. (b)	48. (c)	49. (c)	50. (a)

51. (a)	52. (b)	53. (b)	54. (a)	55. (a)
56. (d)	57. (c)	58. (d)	59. (a)	60. (c)
61. (a)	62. (a)	63. (a)	64. (c)	65. (a)
66. (b)	67. (c)	68. (c)	69. (a)	70. (c)
71. (c)	72. (c)	73. (c)	74. (d)	75. (a)
76. (d)	77. (d)	78. (b)	79. (d)	80. (d)
81. (d)	82. (d)	83. (c)	84. (b)	85. (d)
86. (a)	87. (b)	88. (b)	89. (b)	90. (b)
91. (a)	92. (c)	93. (b)	94. (c)	*95. (b)*
96. (a)	97. (d)	98. (d)	99. (d)	100. (a)
101. (a)	102. (b)	*103. (c)*	104. (a)	105. (a)
106. (b)	107. (b)	108. (a)	109. (a)	110. (d)
111. (c)	112. (b)	113. (c)	114. (b)	115. (b)
116. (a)	117. (b)	118. (d)	119. (b)	120. (a)
121. (d)	122. (b)	123. (c)	124. (d)	125. (b)
126. (b)	127. (d)	128. (c)	129. (b)	130. (a)
131. (a)	132. (c)	133. (a)	134. (b)	135. (b)
136. (b)	137. (b)	138. (a)	139. (d)	140. (c)
141. (b)	142. (a)	143. (a)	144. (c)	145. (a)
146. (a)	147. (c)	148. (a)	149. (a)	150. (c)
151. (d)	152. (b)	153. (d)	154. (d)	155. (a)
156. (a)	157. (c)	158. (b)	159. (b)	160. (b)
161. (c)	162. (d)	163. (a)	164. (a)	165. (c)
166. (d)	168. (d)	169. (c)	170. (a)	171. (a)
172. (a)	173. (a)	174. (c)	175. (a)	176. (a)
177. (d)	178. (d)	179. (b)	180. (b)	181. (c)
182. (b)	183. (a)	184. (a)	185. (c)	186. (b)
187. (b)	188. (c)	189. (b)	190. (c)	191. (d)
192. (c)	193. (c)	194. (a)	195. (b)	196. (c)
197. (c)	*198. (a)*	199. (a)	200. (a)	201. (c)
202. (d)	203. (b)	204. (c)	205. (a)	*206. (b)*
207. (d)	208. (a)	209. (a)	210. (d)	211. (c)
212. (b)	213. (b)	214. *(d)*	215. (d)	216. (b)
217. (b)	218. (b)	219. (c)	220. (c)	221. (a)
222. (a)	223. (d)	224. (a)	225. (a)	226. (d)
227. (b)	228. (a)	229. (a)	230. (b)	231. (a)
232. (a)	233. (b)	234. (b)	235. (b)	236. (c)
237. (c)	238. (c)	239. (c)	240. (d)	241. (c)
242. (d)	243. (b)	244. (c)	245. (a)	246. (a)
247. (d)	248. (d)	249. (d)	250. (c)	251. (c)
252. (a)	253. (c)	254. (d)	255. (d)	256. (c)
257. (a)	258. (c)	259. (b)	260. (a)	261. (b)

262. (a)	263. (a)	264. (b)	265. (b)	266. (c)
267. (a)	268. (c)	269. (b)	270. (d)	271. (c)
272. (b)	273. (a)	274. (d)	275. (a)	276. (d)
277. (a)	278. (c)	279. (b)	280. (a)	281. (b)
282. (a)	283. (d)	284. (a)	285. (a)	286. (b)
287. (a)	288. (c)	289. (b)	290. (c)	291. (a)
292. (d)	293. (d)	294. (c)	295. (b)	296. (c)
297. (c)	298. (b)	299. (a)	300. (c)	301. (c)
302. (b)	303. (b)	304. (c)	305. (b)	306. (c)
307. (b)	308. (b)	309. (b)	310. (b)	311. (a)
312. (a)	313. (a)	314. (c)	315. (a)	316. (d)
317. (a)	318. (c)	319. (d)	320. (a)	321. (d)
322. (a)	323. (d)	324. (b)	325. (a)	326. (d)
327. (d)	328. (a)	329. (acdb)	330. (dcab)	331. (bcad)
332. (d)	333. (a)	334. (a)	335. (b)	336. (d)
337. (c)	338. (d)	339. (d)	340. (a)	341. (c)
342. (c)	343. (a)	344. (c)	345. (b)	346. (c)
347. (a)	348. (a)	349. (b)	350. (d)	351. (d)
352. (a)	353. (d)	354. (d)	355. (b)	356. (b)
357. (b)	358. (c)	351. (d)	352. (a)	353. (d)
354. (d)	355. (b)	356. (b)	357. (b)	358. (c)
359. (d)	360. (d)	361. (d)	362. (b)	363. (d)
364. (c)	365. (a)	366. (a)	367. (c)	368. (d)
369. (c)	370. (a)	371. (b)	372. (b)	373. (c)
374. (b)	375. (c)	376. (a)	377. (c)	378. (a)
379. (d)	380. (c)	381. (c)	382. (c)	383. (a)
384. (c)	385. (d)	386. (c)	387. (d)	388. (c)
389. (d)	390. (b)	391. (c)	392. (c)	393. (a)
394. (d)	395. (b)	396. (b)	397. (d)	398. (a)
399. (c)	400. (a)	401. (c)	402. (c)	403. (a)
404. (a)	405. (a)	406. (d)	407. (c)	408. (a)
409. (b)	410. (b)	411. (b)	412. (c)	413. (d)
414. (a)	415. (b)	416. (b)	417. (b)	418. (c)
419. (c)	420. (b)	421. (d)	422. (d)	423. (c)
424. (c)	425. (c)	426. (a)	427. (c)	428. (a)
429. (a)	430. (b)	431. (c)	432. (d)	433. (c)
434. (b)	435. (c)	436. (b)	437. (b)	438. (d)
439. (c)	440. (a)	441. (a)	442. (c)	443. (a)
444. (a)	445. (c)	446. (c)	447. (d)	448. (a)
449. (b)	450. (a)	451. (a)	452. (c)	453. (c)
454. (c)	455. (c)	456. (b)	457. (b)	458. (a)

459. (a)	460. (c)	461. (c)	462. (c)	463. (b)
464. (b)	465. (b)	466. (c)	467. (a)	468. (b)
469. (a)	470. (c)	471. (a)	472. (a)	473. (a)
474. (a)	475. (d)	476. (c)	477. (d)	478. (b)
479. (c)	480. (d)	481. (b)	482. (b)	483. (c)
484. (b)	485. (c)	486. (c)	487. (b)	488. (b)
489. (c)	490. (a)	491. (b)	492. (b)	493. (c)
494. (b)	495. (a)	496. (d)	497. (b)	498. (c)
499. (a)	500. (c)	501. (d)	502. (d)	503. (c)
504. (c)	505. (a)	506. (b)	507. (a)	508. (d)
509. (a)	510. (d)	511. (c)	512. (c)	513. (a)
514. (d)	515. (c)	516. (b)	517. (a)	518. (d)

Matching the list -I with list -II and select, the correct answer using the given codes:

	List-I		List-II		Codes
519.	**Flood tolerant Categories**		**Tree species**		**Codes**
A.	Very high Mortality	(1)	*Rhododendron*	(a)	A-4,B-2,C-1,D-3
B.	Lower Mortality	(2)	*Rubus procerus* spp.	(b)	A-2,B-1,C-3,D-4
C.	Low mortality	(3)	*Malus* spp.	(c)	A-1,B-3,C-3,D-2
D.	No mortality	(4)	*Prunus laurocerasus*	(d)	A-3,B-4,C-2,D-1
520.	**Plants**		**Survival of root**		**Codes**
A.	Cotton	(1)	7-9 hrs	(a)	A-4,B-2,C-1,D-3
B.	Soybean	(2)	0.5-5 hrs	(b)	A-2,B-1,C-3,D-4
C.	Field pea	(3)	3 days	(c)	A-1,B-3,C-3,D-2
D.	Tomato	(4)	0.5-3.0 hrs	(d)	A-3,B-4,C-2,D-1
521.	**Growth stage of Crops**		**Optimum Temperature**		**Codes**
A.	Maize seedling	(1)	16-21°C	(a)	A-4,B-2,C-1,D-3
B.	Maize root growth	(2)	24°C	(b)	A-2,B-1,C-3,D-4
C.	Field pea	(3)	7-10°C	(c)	A-1,B-3,C-3,D-2
D.	Tomato	(4)	25°C	(d)	A-3B-4,C-2,D-1
522.	**Height of soil Water column (cms)**		**Soil Water Potential**		**Codes**
A.	10.2.	(1)	-15	(a)	A-2,B-1,C-3,D-4
B.	102	(2)	-0.01	(b)	A-1,B-3,C-4 D-2
C.	15,300	(3)	-10.0	(c)	A-3,B-4,C-2,D-1
D.	12000	(4)	-0.001	(d)	A-3,B-4,C-1,D-2
523.	**Soil Minerals**		**Chemical Composition**		**Codes**
A.	Gibbsite	(1)	Iron and Aluminium silicates	(a)	A-4,B-1,C-3,D-4
B.	Feldspar	(2)	Hydrous iron oxides	(b)	A-1,B-3,C-4 D-2
C.	Micas	(3)	Hydrous oxides Aluminium	(c)	A-3,B-4,C-2,D-1
D.	Goethite	(4)	Aluminimosilictes	(d)	A-2 B-4,C-1,D-3

	List-I		List-II	Codes
524.	**Essential elements**		**Quantities in soil**	**Codes**
A.	N	(1)	15-40 kg/ha	(a) A-4,B-2,C-1,D-3
B.	P	(2)	0.1-0.2 kg/ha	(b) A-2,B-1,C-3D-4
C.	K	(3)	140-280 kg/ha	(c) A-1,B-3,C-4,D-1
D.	Ca	(4)	5-20 kg/ha	(d) A-2,B-3,C-1,D-4

525.	**Dominant Minerals**		**Sedimentary Rocks**	**Codes**
A.	Lime stone	(1)	Clays	(a) A-3,B-1,C-4,D-2
B.	Sand stone	(2)	$CaCO_3$(Calcite)	(b) A-1,B-4,C-2 D-3
C.	Shale	(3)	SiO_2(quartz)	(c) A-4,B-2,C-3,D-1
D.	Conglomerate	(4)	Variable	(d) A-2,B-3,C-1,D-4

526.	**Primary Mineral**		**Chemical formula**	**Codes**
A.	Muscovite	(1)	$K\ AlSi_3O_8$	(a) A-4 B-3,C-1,D-2
B.	Biotite	(2)	$Ca\ Al_2Si_2O_8$	(b) A-1,B-4,C-2 D-3
C.	Orthoclase	(3)	$K\ Al\ (Mg\ Fe)\ 3Si4O_{10}(OH)_2$	(c) A-4,B-2,C-3D-1
D.	Anorthite	(4)	$K.Al_3Si_3O_{10}(OH)_2$	(d) A-2,B-4,C-3,D-1A.

527.	**Soil Horizon**		**Striking properties**	**Codes**
A.	A-Horizon	(1)	Eluviation of Fe/Al oxides	(a) A-4,B-1,C-2,D-3
B.	E-Horizon	(2)	Illuviation of minerals	(b) A-1,B-2,C-3 D-4
C.	B-Horizon	(3)	Unconsolidated material	(c) A-2,B-3, C-4,D-1
D.	C-Horizon	(4)	Humified organic matter	(d) A-3,B-4,C-1,D-2

528.	**Nomenclature Classification system**		**Categories in soil**	**Codes**
A.	Mollisols	(1)	Great group	(a) A-3,B-4,C-1,D-2
B.	Aquolls	(2)	Sub group	(b) A-4,B-1,C-2 D-3
C.	Agriaquolls	(3)	Order	(c) A-1,B-2,C-3,D-4
D.	Type Agriaquolls	(4)	Suborder	(d) A-2,B-3,C-4,D-1

529.	**Associated Rhizobium**		**Soil Water Potential**	**Codes**
A.	Organic Humus	(1)	Alfalfa	(a) A-1,B-2,C-4,D-3
B.	Vermiculite	(2)	Soybeans	(b) A-3,B-1,C-2 D-4
C.	Kaolinite	(3)	Lentil	(c) A-4,B-3,C-1,D-2
D.	Hydrous oxide	(4)	Runner bean	(d) A-2,B-4,C-3,D-1

	List-I		List-II	Codes
530.	**Soil Colloids**		**Maximum CEC**	**Codes**
A.	Organic humus	(1)	120 cmol/kg	(a) A-3,B-1,C-4,D-2
B.	Vermiculite	(2)	3 cmol/kg	(b) A-1,B-4,C-2 D-3
C.	Kaolinite	(3)	300 cmol/kg	(c) A-4,B-2,C-3,D-1
D.	Hydrous oxide	(4)	5 cmol/kg	(d) A-2,B-3,C-1,D-4A.
531.	**Organic matter**		**C:N ratio**	**Codes**
A.	Legumes	(1)	4:1 to 9:1	(a) A-1,B-4,C-3,D-1
B.	Straw residues	(2)	400:1	(b) A-2,B-1,C-4 D-3
C.	Saw dust	(3)	100:1	(c) A-4,B-3,C-2,D-1
D.	Microbe bodies	(4)	20:1 to 30:1	(d) A-3,B-2,C-1,D-4
532.	**Amendments (in tones)**		**Equivalent to 1 tones of S**	**Codes**
A.	Gypsum	(1)	8.69	(a) A-3,B-2,C-4,D-1
B.	Lime stone	(2)	5.38	(b) A-1,B-3,C-2 D-4
C.	Iron sulphate	(3)	3.06	(c) A-4,B-1,C-3,D-2
D.	Sulphuric acid	(4)	3.13	(d) A-2,B-4,C-1,D-3
533.	**pH of saturated $CaCO_3$ solution**		**Solubility (meq/L)**	**Codes**
A.	6.21 pH	(1)	1.1	(a) A-3,B-4,C-1,D-2
B.	7.12 pH	(2)	0.82	(b) A-4,B-1,C-2 D-3
C.	8.60 pH	(3)	19.3	(c) A-1,B-2,C-3,D-4
D.	9.20 pH	(4)	7.1	(d) A-2,B-3,C-4,D-1
534.	**Conversion Form**		**Conversion Factor**	**Codes**
A.	Ca to CaO	(1)	2.50	(a) A-1B-3,C-2,D-4
B.	Ca to Ca $(OH)_2$	(2)	1.40	(b) A-4,B-1,C-3 D-2
C.	Ca to $CaCO_3$	(3)	1.67	(c) A-2,B-4C-1,D-3
D.	Mg to MgO	(4)	1.85	(d) A-3,B-2,C-4,D-1
535.	**Sodium uptake**		**Crops**	**Codes**
A.	High	(1)	Buck wheat	(a) A-2,B-1,C-4,D-3
B.	Medium	(2)	Millet	(b) A-1,B-4,C-3 D-2
C.	Low	(3)	Cabbage	(c) A-4,B-3,C-2,D-1
D.	Very low	(4)	Spinach	(d) A-3,B-2,C-1,D-4
536.	**Sodic soils**		**Exchangeable Sodium %**	**Codes**
A.	Slightly sodic	(1)	15-20	(a) A-1,B-2,C-3,D-4
B.	Moderately sodic	(2)	20-30	(b) A-2,B-3,C-4 D-1
C.	Very sodic	(3)	>30	(c) A-3,B-4,C-1,D-2
D.	Extremely sodic	(4)	7-15	(d) A-4,B-1,C-2,D-3

	List-I		List-II		Codes
537.	**Crops**		**Sodium uptake**		**Codes**
A.	Sugar beat	(1)	Medium	(a)	A-3,B-1,C-4,D-2
B.	Potato	(2)	Very low	(b)	A-1,B-4,C-2 D-3
C.	Barley	(3)	High	(c)	A-4,B-2,C-3,D-1
D.	Maize	(4)	Low	(d)	A-2,B-3,C-1,D-4
538.	**Zn deficiency disease**		**Crops**		**Codes**
A.	White bud	(1)	Rice-paddy	(a)	A-3,B-4,C-2,D-1
B.	Little leaf	(2)	Citrus crop	(b)	A-4,B-2,C-1 D-3
C.	Mottle leaf	(3)	Maize and Sorghum	(c)	A-2,B-1,C-3,D-4
D.	Khaira	(4)	Cotton	(d)	A-1,B-3,C-4,D-1
539.	**Clay mineral**		**CEC (meq/100 g)**		**Codes**
A.	Mineral colloids 1::1	(1)	20-40	(a)	A-3,B-4,C-1,D-2
B.	Mineral colloids 2::1	(2)	100-300	(b)	A-4,B-1,C-2 D-3
C.	Chlorite 2::1::1	(3)	1-10	(c)	A-1,B-2,C-3,D-4
D.	Organic colloids	(4)	80-150	(d)	A-2,B-3,C-4,D-1
540.	**Soil Test rating**		**Fertility Index**		**Codes**
A.	Very low	(1)	25-50	(a)	A-1,B-2,C-3,D-4
B.	Low	(2)	50-100	(b)	A-2,B-3,C-4 D-1
C.	Medium	(3)	0-10	(c)	A-3,B-4,C-1,D-2
D.	High	(4)	10-25	(d)	A-4,B-1,C-2,D-3
541.	**Symptoms**		**Causes**		**Codes**
A.	Khaira disease	(1)	P-deficiency	(a)	A-4,B-2,C-1,D-3
B.	Browning in Veg.	(2)	Mo-deficiency	(b)	A-2,B-1,C-3 D-4
C.	Purple in Maize	(3)	K-deficiency	(c)	A-1,B-3,C-4,D-2
D.	Firing in Tobacco	(4)	Zn-deficiency	(d)	A-3,B-4C-2,D-1
542.	**Fertilizers**		**Equivalent acidity**		**Codes**
A.	Urea	(1)	148	(a)	A-4,B-3,C-2,D-1
B.	Ammonium Sulphate	(2)	128	(b)	A-3,B-2,C-1 D-4
C.	Ammonium Chloride	(3)	110	(c)	A-2,B-1,C-4,D-3
D.	Anhydrous NH_3	(4)	84	(d)	A-1,B-4,C-3,D-2

	List-I		List-II		Codes
543.	**Name of fertilizers**		**Types of fertilizers**		**Codes**
A.	Ammonium Sulphate	(1)	Biofertilizer	(a)	A-2,B-4,C-1,D-3
B.	DAP	(2)	Straight fertilizers	(b)	A-4,B-1,C-3 D-2
C.	Algae	(3)	Organic manure	(c)	A-1,B-3,C-2,D-4
D.	FYM	(4)	Complex fertilizer	(d)	A-3,B-2,C-4,D-1

544.	**Formative elements**		**Connotation**		**Codes**
A.	Acr	(1)	Cool	(a)	A-4,B-1,C-2,D-3
B.	agr	(2)	Cold	(b)	A-3,B-4,C-1 D-2
C.	bor	(3)	Extreme weather	(c)	A-1,B-2,C-3,D-4
D.	cry	(4)	agric horizon	(d)	A-2,B-3,C-4,D-1

545.	**Material**		**Equivalent to 1 tones of gypsum**		**Codes**
A.	Sulphuric acid	(1)	0.75	(a)	A-4,B-2,C-1,D-3
B.	Sulphur	(2)	0.18	(b)	A-2,B-3,C-4 D-1
C.	Lime sulphur	(3)	1.62	(c)	A-2,B-4,C-3,D-1
D.	Iron sulphate	(4)	0.57	(d)	A-2,B-1,C-4,D-3

546.	**Soil Order**		**Major characteristics**		**Codes**
A.	Alfisols	(1)	Low Base Saturation	(a)	A-3,B-1,C-4,D-2
B.	Histosols	(2)	>30% OM	(b)	A-1,B-4,C-2 D-3
C.	Ultisols	(3)	Argillic/ nitric horizon	(c)	A-4,B-2,C-3,D-1
D.	Vertisols	(4)	Deep crack, when dry	(d)	A-2,B-3,C-1,D-4

547.	**Soil Order**		**Organic Matter %**		**Codes**
A.	Aridisols	(1)	2.0	(a)	A-4,B-2,C-1,D-3
B.	Histosols	(2)	4.0	(b)	A-3,B-4,C-2 D-1
C.	Mollisols	(3)	1.0	(c)	A-2,B-1,C-3,D-4
D.	Vertsols	(4)	80.0	(d)	A-1,B-3,C-4,D-2

548.	**Liming Material**		**$CaCO_3$ equivalent (%)**		**Codes**
A.	Lime stone	(1)	180.0	(a)	A-2,B-3,C-4,D-1
B.	Dolomite	(2)	109	(b)	A-2,B-3,C-2 D-4
C.	Burnt Lime	(3)	100	(c)	A-3,B-2,C-1,D-4
D.	Hydrated Lime	(4)	136.0	(d)	A-2,B-4,C-3,D-1

	List-I	List-II	Codes
549.	**Soil Constitutes**	**Specific Gravity**	**Codes**
A.	Lime stone	(1) 2.5-2.8	(a) A-1,B-2,C-3,D-4
B.	Mica	(2) 1-1.7	(b) A-4,B-1,C-2 D-3
C.	Quartz	(3) 2.8-3.2	(c) A-4,B-3,C-1,D-2
A.	Humus	(4) 3.4-4.0	(d) A-2,B-3,C-4,D-1
550.	**Soil Types**	**Organic Matter (%)**	**Codes**
A.	Minerals soils	(1) 15-30	(a) A-1,B-2,C-3,D-4
B.	Humic soils	(2) Less than 2	(b) A-4,B-1,C-2,D-3
C.	Anmoor peat	(3) More than 30	(c) A-3,B-4,C-1,D-2
D.	Peat soils	(4) 2-15	(d) A-2,B-4,C-1,D-3
551.	**Crops**	**Optimum pH range**	**Codes**
A.	Lucerne	(1) 4.0-7.0	(a) A-1,B-2,C-3,D-4
B.	Wheat	(2) 6.5-7.4	(b) A-2,B-4,C-1 D-3
C.	Oats	(3) 4.1-5.5	(c) A-3,B-4,C-1,D-2
D.	Lupines	(4) 4.1-7.4	(d) A-2,B-3,C-4,D-1
552.	**Plants**	**Boron content (ppm)**	**Codes**
A.	Organic humus	(1) 94.0	(a) A-3,B-2,C-4,D-1
B.	Vermiculite	(2) 22.0	(b) A-4,B-1,C-2 D-3
C.	Kaolinite	(3) 2.3	(c) A-3,B-4,C-1,D-2
D.	Hydrous oxide	(4) 49.0	(d) A-2,B-3,C-4,D-1
553.	**Ecosystems**	**Biological fixation (kg /N /ha /yr)**	**Codes**
A.	Arable Land	(1) 73-865	(a) A-1,B-2,C-3,D-4
B.	Non-legumes	(2) 58-594	(b) A-2,B-4,C-1 D-3
C.	Grass legume	(3) 7-28	(c) A-3,B-4,C-1,D-2
D.	Forest	(4) 7-114	(d) A-2,B-3,C-4,D-1
554.	**Clays Minerals**	**AEC (meq/100g)**	**Codes**
A.	Bentonite	(1) 7.7	(a) A-1,B-2,C-3,D-4
B.	Illite	(2) 15	(b) A-4,B-1,C-2 D-3
C.	Vermiculite	(3) 9	(c) A-3,B-4,C-1,D-2
D.	Montmorillonite	(4) 43	(d) A-2,B-3,C-4,D-1
555.	**Chelates**	**Chemical formula**	**Codes**
A.	EDTA	(1) $C_{14}H_{23}O_{10}N_3$	(a) A-3,B-1,C-4,D-2
B.	DTPA	(2) $C_{14}H_{22}O_8N_2$	(b) A-1,B-4,C-2 D-3
C.	CDTA	(3) $C_{10}H_{16}O_8N_2$	(c) A-4,B-2,C-3,D-1
D.	EDDHA	(4) $C_{18}H_{20}0_6N_2$	(d) A-2,B-3,C-1,D-4

Answers

519. (a)	520. (a)	521. (c)	522. (d)	523. (a)
524. (a)	525. (d)	526. (a)	527. (a)	528. (a)
529. (d)	530. (a)	536. (d)	537. (a)	538. (a)
539. (a)	540. (c)	541. (a)	542. (a)	543. (a)
544. (b)	545. (a)	546. (c)	547. (b)	548. (c)
549. (c)	550. (d)	551. (b)	552. (a)	553. (c)
554. (d)	555. (a)			

III. FILL IN THE BLANKS

556. The four major layers that make up the earth are the crust —— —————— and ——————

557. The soil forms a thin layer between the —————————— ————--and ——————————

558. In earth with increase in depth, there is a tremendous ———— ——————--in temperature.

559. About ——————————————-percent of the earth's surface covered with water.

560. Lithosphere accounts for —————————— ——percent of the earth's mass.

561. At the sea level atmosphere exerts a pressure of ———— ——————g./cm^2

562. In general, primary minerals are presents in coarse fraction of the soil whereas secondary mineral are dominant in the ———— ————————--fraction of the soil.

563. Soil formation is a ——————.——————

564. Formation of one inch soil needs ——————————

565. Horizon 'A' also called as ——————————

566. Horizon 'B' also called as ——————————

567. Bulk density of general soil for calculation is ——————————

568. Particle density of normal soil is ——————————

569. The particle density of organic matter is ——————————

570. Soil temperature required for optimum crop root development is ——————————

571. Farm yard manure contain——————————

572. Poultry manure contain——————————

573. Triple super phosphate contain ——————————

574. Aqueous ammonia contain ——————————

575. *Rhizobium* biofertilizer can fix ——————————————————-per year.

576. Thiourea contain ——————————————————————-

577. ——————————————————-of *Rhizobium,* culture is required to treat 1 kg seed.

578. BGA can add up to about ———————————————-N/ha to rice field.

579. Most minerals with a metallic luster are good conductors of ————————and ————.

580. Mineral may be classified as light or heavy depending upon their——————————-.

581. The ——————the volume occupied by a given number of oxygen is the more stable is the mineral.

582. ——————————————————————is a complex ring borosilicate of aluminium.

583. Pyrite contains——————————% iron ——————————————% sulphur.

584. ——————————————————————————is non-plastic clay.

585. Chemical composition of olivine may be represented by formulae———————————

586. *Azotobacter* and *Azospirillum* secrete antibiotics which act as pesticides so biofertilizers also acts as ——————————————————————

587. Legumes fails to develop nodules when pH less than——————————————-

588. *Azospirillum* is used for ——————————————————————

589. Pahala blight of sugarcane is used by the deficiency of ———————————————-

590. The chemical composition of gypsum is ——————————————————

591. Boron availability at pH ———————————————————————-

592. The optimum pH range for tea and rice is ——————————————————-

593. The soil consist of four major component i.e.——————organic matter, soil air, and ————

594. ——————————————- refers to the relative percentage of sand, silt, clay in a soil.

595. The water holding capacity of clay is ——————————

596. ——————————
———is an artificially formed soil mass.

597. Puddling of the soil is generally beneficial to the production of —
——————————

598. Particle density is also called as ——————————

599. Particle density of normal soil is ——————————

600. —————— refers to that percentage of soil volume which is occupied by pore spaces.

601. The bulk density of sand soil is ——————————

602. Biological agencies accelerate physical and ——————————

603. The excavation and destructive action of water is called — ———
——————————

604. Dry soil has a specific gravity of —————— ——————

605. The reaction of a compound with carbonic acid is called ———
——————————

606. The mineral sediments deposited in fresh water are called ———
——————————

607. When water freezes, it expends to about ——————————
————percent of its volume

608. Oxidation of ferrous oxide leads to formation of mineral— ———
——————————

609. Any soil properties is a function of the collective effects of ———
——————————

610. In high rainfall regions the clay content——————————
——————————with depth.

611. Formation of glei layer is a characteristics of process called ———
——————————

612. The origin material from which soil develops is called —— ——
——————————

613. A soil series may be classified into types on the basis of ———
——————————

614. Genetic system of soil classification is based entirely on - ———
——————————

615. Diagnostic horizons found at the surface of the soil are called—
————————————————

616. Man made horizon formed during use of soil by people for long time is called————————

617. Soil of the order ————————————————————are dark coloured soils of grassland.

618. Formation of soil order———————————is common in hot areas having stagnant marshes and swamps.

619. ————————————————————is a list of defined mapping units with their symbols.

620. The black colour of soils is due to the presence of ————————compounds/ clay humus complexes.

621. In general, red soil are derived from ———————————— and ————————————

622. In reconnaissance soil survey, the soil units delineated are of a higher categories i.e. ————————————————.

623. In————————————————surveys topsheets of the survey used as base maps.

624. ————————————————————————— is used to examine aerial photographs.

625. Boron availability at pH——————————————————— ————————————

626. Sandy soils have large number of ——————————————————————

627. Clayey soils have a larger number of ——————————————————————

628. The quality of adhesion to other object is called———————————————————

629. The weight per unit volume of the solid portion of soil is called—
————————————

630. The oven dry weight of a unit volume of soil inclusive of pore space is called ————————

631. Black colour and dark grey colour of soil is due to ————————————————————

632. Red colour of soil is due to ——————————————————— ————————————

633. White colour of soil is due to ————————————————————— ————————————

634. Capillary water is therefore known as ———————————————————————

635. The water beyond the field capacity stage is called to be ——————————————

636. Available water that water which lies between —————— ——and——————————

637. Water at wilting coefficient is held with a force of ——— ——— ——————————

638. The movement of water through a column of soil is called ——— ——————————

639. —————————————————— ————are poor in available Ca and Mg

640. Phosphorus availability is highest when pH is between—— —— ——————————

641. The chemical composition of dolomite limestone is ——— ——— ——————————

642. Highly acid tolerant crop is —————————————— — ——————————

643. The addition of nitrogen in the soil by bacteria is called — —— ——————————

644. ——————————————is an aerobic nitrogen fixing bacteria and thrives well in

645. Mutually beneficial association between certain fungi and roots of higher plants called——————.

646. Phosphorus is absorbed by plants in the form of ——————— — ——————————

647. The excess of N, K and Ca may reduce —————— ——————toxicity.

648. Basic slag classified in the —————————— ——————phosphorus

649. The land capability class further divided into—————— depending upon the kindof limitation for its use.

650. Land use for Class VIII land is restricted to —————— ——————————

Answers

556. (the mantle the outer, the inner core)

557. (atmosphere and crust) **558.** (increase)

559. (93.06) **560.** (71)

561. (1033.41) **562.** (finer fraction)

563. (slow process) **564.** (800-1000 years)

565. (Eluviation horizon) **566.** (illuviation horizon)

567. (1.33 g/cm^3) **568.** (2.65 g/m^3)
569. (1.1 -1.4 g/m^3) **570.** (15-45°C)
571. (0.5% N) **572.** (3.03 %N)
573. (46-48 % P_2O_5) **574.** (80% N)
575. (50-200 kg N/ha) **576.** (36.8% N)
577. (20 g) **578.** (20-25 kg)
579. (heat & electricity) **580.** (specific gravity)
581. (smaller) **582.** (quartz)
583. (46 and 53) **584.** (fuller's earth)
585. ($(Mg, Fe)_2 SiO_4$) *586. (bio-pesticide)*
587. (pH 5.0) **588.** (Sorghum)
589. (Mn) **590.** ($CaSO_4.2H_2O$)
591. (5.0-7.0) **592.** (4 to 6)
593. (mineral matter, soil water)
594. (soil texture) **595.** (very high)
596. (C lod) **597.** (Rice)
598. (True density) **599.** (2.65 g/cm^3)
600. (Soil porosity) **601.** (1.60 g/cm^3)
602. (Chemical) **603.** (denudation)
604. (2.65) **605.** (Carbonation)
606. (Lacustrine) **607.** (9)
608. (hematite) **609.** (Soil forming factor)
610. (increase) **611.** (gleilayer)
612. (Parent material) **613.** (Soil texture)
614. (Soil genesis) **615.** (Epipedons)
616. (Anthropic epipedon) **617.** (Mollisols)
618. (Histosols) **619.** (Legend)
620. (Titaniferous magnetite) **621.** (granite and gneisses)
622. (soil association) **623.** (reconnaissance)
624. (Video Image Analysis) **625.** (5.0-7.0)
626. (macro pores) **627.** (micro pores)
628. (stickiness) **629.** (true density)
630. (bulk density) **631.** (organic matter)
632. (ferric oxide) **633.** (silica & lime)
634. (available water) **635.** (super available water)
636. (wilting coefficient, field capacity)
637. (15 atmosphere) **638.** (Percolation)

639. (Acid soil) **640.** (6.5 to 7.5)
641. (Ca Mg $(CO_2)_2$ **642.** (sweat potato)
643. (bacterial fertilizer) **644.** (Clostridium)
645. (acid soil) **646.** ($H_2PO_4^-$ & $HP)_4^-$)
647. (Boron) **648.** (Boron)
649. (natural vegetation of forests& grasses)
650. (not recommended)

IV. FREQUENTLY ASKED QUESTION (FAQ)

651. Indicates the purity of the colour

Ans. Chroma

652. Mechanical analysis is also called

Ans. Particle size analysis

653. Sand is chemically

Ans. inactive

654. When the faces and edges in blocky structure are rounded, the structure called.

Ans. Sub angular blocky

655. bulk density of soil is expressed

Ans. Mg / m^3

656. If a soil contains 10 ppm of available phosphorus, this is equivalent to

Ans. 20 kg /ha

657. The soil water containing nutrients ions is known as

Ans. Soil solution

658. Capillary water is held in microspores at tensions between

Ans. 0.1 to 31 bars

659. Neutral scattering method is not suitable.

Ans. In organic soil

660. The unit term is equal to 1 atmosphere

Ans. Bar

661. Soil air contains more carbon dioxide than the atmosphere air

Ans. 5 to 10 times

662. Thermal capacity is expressed as

Ans. Cal / g / °C

663. Wheat grows best at temperature

Ans. 24 °C

664. The specific heat of humus

Ans. 0.443 Cal / g / °C

665. Organic soil colloids

Ans. Humus

666. Allophane have a chemical composition

Ans. 1::1

667. The CEC of vermiculite

Ans. 150 cmol / kg of soil

668. Capacity of a soil to exchange cations is the best single index

Ans. Soil fertility

669. The product of concentration of H^+ and OH^- is called

Ans. Ionic product of water

670. If water soluble boron is more than 1 ppm

Ans. Unsuitable for irrigation

671. Salinity is measured in term of

Ans. Electrical conductivity (EC)

672. Any soil having pH greater than 7.0

Ans. Alkaline soil

673. The neutralizing value of $CaCO_3$

Ans. 100 %

674. The neutralizing value of Burnt lime

Ans. 179%

675. The neutralizing value of Slaked lime

Ans. 136%

676. The neutralizing value of Dolomite

Ans. 109%

677. Lime reduces excessive uptake of nutrient

Ans. Potassium

678. The available nitrate and ammonium form is the % of total soil nitrogen

Ans. 1-2 %

679. Two form of nitrogen available to plant

Ans. Nitrate and Ammonium

680. Mineralization of organic N involve two reaction

Ans. Aminization and ammonification

681. Ammonium sulphate nitrate contains sulphur

Ans. 12 % sulphur

682. Organic phosphorus represents the total P % in soils.

Ans. 50 %

683. The C:N:S ratio of soil organic matter

Ans. 125:10:1.2 (Approximately)

684. Single super phosphate contains approx. percent of sulphur

Ans. 12.0 %

685. Elemental sulphur contains % of sulphur

Ans. 100 %

686. Essential constituent of all green plant

Ans. Magnesium

687. The most prominent boron mineral in soil

Ans. Tourmaline

688. The form Boron occurs at pH greater than 8.5

Ans. $B(OH)_4$

689. iron has the tendency within the plant

Ans. Immobile

690. Liming of acids soil increase availability of phosphorus due to

Ans. Anion exchange

691. Those micro-organism which grow at temperature below 10 ^{o}C

Ans. Psychrophiles

692. Bacteria capable of developing both in the presence and absence of oxygen

Ans. Facultative anaerobic

693. Fungi which attack lignin, cellulose and related compounds

Ans. Basidiomycetes

694. When two micro-organism behave entirely independently

Ans. Neutralism

695. The C:N ratio of humus is (approx.)

Ans. 1:10

696. Which enzyme breaks down urea to water, CO_2 and NH^+.

Ans. Urease

697. The P-containing lipids are also called

Ans. Phospho-lipid

698. The basic component of lipids

Ans. Glysol

699. N content of most protein is about

Ans. 16 %

700. Starch is the polymer of

Ans. Glucose

701. How much tons of nitrogen available in atmosphere of each hectare of land

Ans. 80,000 tons

702. In water logged paddy soil, which are important source of nitrogen.

Ans. Blue Green Algae (BGA)

703. Tryptophan produce in the roots of the legume, in which is converted by Rhizobia

Ans. Iodole Acetic Acid (IAA)

704. The chemical which damage small mammals

Ans. Rodenticides

705. Fumigation is done for the control of

Ans. Nematocides

706. Phosphate fertilizers contains on an average % of cadmium /g of fertilizers.

Ans 7 mgm

707. Lead occurs in soil as

Ans. PbO_2

708. Excess of which chemical, in drinking water causes dental and skeletal disorder

Ans. Fluorine

709. Arsenic in soil is present in anionic form

Ans. $H_2AsO_4^-$

710. Insoluble form of mercury are converted to a soluble inorganic form called

Ans. Methyl mercury

711. Nitrate levels in drinking water contaminated with bacteria cause

Ans. Blue baby disease (Methaaemoglobmemia)

712. Nitrate levels in drinking water are considered a human health hazard

Ans. 10 mg per litre

713. Excess of potassium in grass accompanied with higher applications of nitrogen may leads in cattle

Ans. Grass tetany (Hypomagnesemia)

714. Direct source of Ca for plant growth

Ans. Gypsum and phosphogypsum

715. Fruit end rot in tomato caused by

Ans. Deficiency of Ca

716. The book 'Soil Fertility and fertilizers' was written by

Ans. Tisdale *et. al.*

717. Who gave the statement 'Water was the sole nutrient of plants'

Ans. JB Helmont

718. "Annals of Agriculture" a book was written by

Ans. Arthur Young

719. Which element considered as energy currency for the plant?

Ans. Potassium

720. Fe deficiency is mostly seen in crop growing on

Ans. Calcareous or alkaline soils

721. Crinkle leaf of cotton is caused by

Ans. Mn toxicity

722. The negative charge on 2::1 clays is

Ans. pH dependent and pH independent

723. One acre of soil to a depth of 6 inch weight about

Ans. 2×10^6 lb

724. For nitrification, the optimum pH is

Ans. Around 8.5

725. *Nitrosomonas* and *Nitrobacter* are an obligate

Ans. Autotrophic bacteria

726. Maximum nitrification takes place at

Ans. 20 % Oxygen

727. Biuret in urea applied as foliar spray in citrus fruits should not be more than

Ans. 0.25 %

728. At what pH, there are approx. equal amounts of $H_2PO_4^-$ and HPO_3^{2-}

Ans. pH 7.2

729. What is the porosity percentage if the particle density of soil is 2.6 and bulk density is 1.3

Ans. 50 %

730. Height of rise of water in a capillary tube is directly proportional to

Ans. Surface tension of water

731. Hill soils are generally

Ans. Acidic soil

732. The diagnostic surface horizons are called

Ans. Epipedons

733. Available water holding capacity is maximum in

Ans. Clay loam

734. Concentration of which element is highest in soil.

Ans. Oxygen

735. Specific heat is higher in

Ans. Humus

736. Podzolization is a process of accumulation in soil

Ans. Silica

737. Laterization is a process of accumulation in soil

Ans. Iron and aluminium oxides

738. Rocks containing more than 65 % SiO_2 are

Ans. Acidic

739. Critical limit of available phosphorus in soil for wheat crop by Olsen's method is

Ans. 10 kg /ha

740. In which soil, Zn availability to crop is higher

Ans. Acidic soil

741. The sequence of textural classes of soil with increasing order of fineness of texture is

Ans. LS, SL, L, SCL, CL

742. Humic acid isolated from soils is

Ans. Aromatic with high molecular weight

743. In which oil cake, the highest nitrogen content is

Ans. Ground nut

744. From Q / I relationship described by Beckett (1964), a high P B C^k indicates

Ans. The availability of K in the soil is excellent

745. Sodium Adsorption Ratio (SAR) can be calculated by using of formula

Ans. SAR= Na^+ / "Ca $^{2+}$ + Mg^{2+}/ 2

746. Lowest categories of soil classification

Ans. Series

747. Which water potential in soil is always positive

Ans. Gravitational potential

748. 'White alkali' soils are called

Ans. Saline soils

749. The smallest volume that can be called 'Soil' is

Ans. Pedon

750. Soils having 200 kg available K_2O / ha will be classed as

Ans. Medium

751. Sugar beat is an indicator plant for

Ans. Sodium

752. On functional basis, Fe, Zn, Mn, and Cu have been classified as

Ans. Catalyst and activators

753. The dominant spectral colour in soil indicates
Ans. Hue
754. The pH of the soil solution determines
Ans. Active acidity
755. The value of ESP in alkali soil is always
Ans. Greater than 15
756. The moisture equivalent was introduced by
Ans. Briggs and McLane (1907)
757. A mature and dry plant tissue contain about protein %
Ans. 1- 15 %
758. Flour apatite is a rich source of
Ans. Potassium
759. Fe concentration used for foliar spray is
Ans. 0.4 %
760. Who developed method for estimation of lime requirement
Ans. Shoemaker *et. al.*(1961)
761. Indicator plant of Cu is
Ans. Wheat
762. Chemical formula of Siderite is
Ans. $FeCO_3$
763. The correct sequence of weathering stages in soil formation is
Ans. Initial"Juvenile"Virile"Senile"Final
764. Hydromorphic soils are evolved by which process
Ans. Gleization
765. Pedoturbation is the process of
Ans. Mixing of the soil
766. Which soil order have highest area in India
Ans. Inceptisols
767. Which soil order have negligible area in India
Ans. Histosols
768. A soil profile having two sola, one above the other is known as
Ans. Bisequem profile
769. Humus forming fungi includes which genera
Ans. *Alternaria* and *Aspergillus*
770. Nutrient index between 1.5 and 2.5 is categorized as
Ans. Medium

771. Fertilizer control order was issued in

Ans. 1957

772. Gapon's equation is widely used equation to describe

Ans. CEC of the soil

773. The value of Van Bemmelen factor is

Ans. 1.724

774. Who is considered as father of soil testing

Ans. Troug

775. Fertility gradient approach was given by

Ans. B. Ramamurthy

776. Which country has largest deposits of K

Ans. Canada

777. Which country have monopoly in P deposits

Ans. Morocco

778. In India, which fertilizer manufactured first

Ans. SSP

779. Harting nets in soil produce by

Ans. Ectomycorrhiza

780. The end product of denitrification process is

Ans. N_2O and N_2

781. Which nitrogenous fertilizer industrially made first

Ans. $CaCN_2$

782. The ultra-micronutrient elements or nanonutreint

Ans. Molybdenum

V. CHOOSE THE CORRECT ANSWER

783. Soil air contents a **higher / lower** concentration of carbon dioxide.

784. Quartz has a **hexagonal / tetragonal** crystal form

785. Olivine has a **trimetric / triclinic** crystal form.

786. Graphite and pyrite have **a metallic / non-metallic** luster.

787. Pyroxene and amphiboles are **heavy / light** mineral

788. Quartzites are **non- foliate / foliate** rocks.

789. Quartz is classified as a **tectosilicate / phyllosilicates**

790. Doubling the velocity of water increase the amount of material carried **32 times / 64 times.**

791. High temperature accelerates the process of weathering especially in **humid / arid regions.**

792. Water is **necessary / not necessary** for chemical weathering.

793. Oxidation involves **removal / addition** of electrons.

794. Iron and manganese impart **red / black** colour to soils.

795. **Solum / Regolith** are the true soil

796. **Podzolization / Laterization** commonly occurs in forests.

797. Swelling of clays is **Argillic Pedoturbation / floral Pedoturbation.**

798. Soil characteristics determined by local conditions in **Intrazonal soils / azonal soils.**

799. Torric moisture regimes occur in **arid / humid regions.**

800. When the mean annual soil temperature is 22°C or higher, the soil temperature regimeis called **hyper thermic / thermic**

801. Vertisols is the common soil order for the **black / red** soil.

802. **Class III / Class II** land is categorized as moderately

803. **Silt / Clay** feels smooth like flour or talcum powder when dry.

804. **Type / Class** refers to the shape of soil aggregates.

805. The bulk density of a soil is always **smaller/ greater** than its particle density.

806. Organic matter is much **lighter/ heavier** than the corresponding volume of mineral matter.

807. In course, textured soils the pores are **non-capillary / capillary.**

808. Hygroscopic water is mostly in **non-liquid/ liquid form.**

809. Free energy also called **potential / kinetic energy.**

810. At moisture equivalent, the negative potential is approximately **0.5 bars / 1 bar.**

811. Excessive tillage **reduces/ increase** filtration.

812. **Diffusion / Mass flow** accounts for most of the gaseous interchange.

813. The diffusion is **inversely / directly proportion** to the square root of the density of the gas.
814. Heat capacity of humus is **less / more** than that of mineral matter.
815. The thermal capacity of sand is **more /less** than clay.
816. Heat flow is **slower / faster** in soil than in the atmosphere.
817. Soil containing more than **15 % / 5 %** clay exhibits plasticity.
818. Muscovite and Biotite are two main members of **2::1 non-expending / 2::1 expending mineral.**
819. Isomorphous substitution occurs significantly in **2::1 / 1::1 type of clays.**
820. Micelles can consider as **pollinations / polycations** because they carry largemembers of negative charges.
821. **Montmorillonite / Kaolinite** dominated soils are batter drained.
822. **Molybdenum / Manganese** become more available with increasing soil pH.
823. At pH 5.5, $\mathbf{H_2PO_4^-}$ **/** $\mathbf{PO_3^-}$ is dominant.
824. At equal base saturation, **montmorillonite / kaolinite** dominant solution will containmore H- ions.
825. Saline soils are in a **flocculated / de flocculated state.**
826. Flow of water and nutrients take place from **lower / higher** concentration towards higher concentration.
827. Brackish water has **chloride / nitrate salts.**
828. Lime **may not be applied / may apply** along with nitrogenous fertilizers.
829. Liming **reduces/ increases** plant uptake of heavy metals.
830. Excess aluminium accumulates in the roots and **reduces / increases** their power of translocating phosphates from the soil.
831. The greater part of nitrogen occurs in **organic / inorganic forms.**
832. Nitrogen deficiency first observed when **older / newer** leaves of plants turn yellow.
833. The primary source of nitrogen is the **atmosphere / soil.**
834. Ammonium sulphate nitrate is **non-hygroscopic /** hygroscopic.
835. Mono and Di-calcium phosphate are **readily available / not readily available** for plants.
836. Ammonium nitrate is **soluble / insoluble** in water.
837. Sulfate is mainly Tran located in an **upward / downward** direction.
838. **Atmosphere / Soil** is most important source of sulphur.
839. Sulphur **enhances / reduces** uptake of phosphorus in plants.
840. Sulphur mineralization is favoured in soils with **low / high** C: S ratio.

841. Application of organic matter leads to **loss /gain** of Ca and Mg from the soil.

842. **Mg / Ca** is an essential constituent of all green plants.

843. **Foliar application / Soil application** of micronutrients is more suitable to avoid elemental interaction.

844. **Zn / Fe** is present in Cytochrome oxidase.

845. Chlorophyll, the green pigment in leaves is a chelate of **Mg / Fe.**

846. Major part of cation exchange capacity of organic matter is due to the presence of **COOH group / Phenolic groups.**

847. Zinc is more strongly adsorb by **magnesite / calcite.**

848. Nitrogen fertilizers that are acid forming will **increase/ decrease** the uptake of zinc.

849. Liming to correct soil acidity will improve **Mo/ Mn** deficiency.

850. **Borax / Boric acid** is soluble in water and therefore easily available to plants.

851. Microorganisms, which can proliferate in higher than 45°C temperature, are **thermophiles / mesophiles.**

852. **Photoautotroph / Chemoautotroph** derive **their** energy from sunlight.

853. Diatoms are a type of **algae / fungi.**

854. Molds, yeast and mushrooms grouped as **fungi / algae.**

855. Azotobacter is a **non-symbiotic / symbiotic** nitrogen-fixing bacteria.

856. Inorganic soils, microbial activity is greater at **deeper layer/ surface.**

857. **Plant tissue / animals** are the main source of soil organic matter.

858. When fresh plant residues added to the soil the C:: N is **wide/ narrow.**

859. **Humic / Non- humic** substances are more resistant to microbial attack.

860. Polysaccharide, polymers, and ployuronides are **Humic / Non-humic** substances.

861. Humus is **acidic / alkaline**.

862. The colloidal surfaces of humus are **negatively / positively charge.**

863. **Bacteria / Fungi** are heavy users of nitrogen.

864. **Twenty-one / eleven amino acids** usually found as protein constituents.

865. **Hemicelluloses/ cellulose** attacked first by microorganism.

866. **Azotobacter / Azospirillum** is capable of utilizing carbon sources in soil.

867. **Azolla / Azotobacter** can be utilizing under stagnant water conditions like rice cultivation.

868. All the *Rhizobial* strains, found in soils, are **beneficial / not beneficial.**

869. In general, **herbicides / fungicides** are more mobile.

870. 2, 4-D is rapidly degraded under **aerobic / an aerobic conditions.**

871. Most of the carbamates are **toxic / not toxic** to earth worn.

872. Herbicide trifuralin is **volatile / nonvolatile**.

873. **Cadmium / lead** is highly soluble.

874. The accumulation of heavy metals is more in the **roots / shoots.**

875. Most of the atmosphere ozone found in the **stratosphere / mesosphere.**

Answers

783. (higher)	**784.** (hexagonal)
785. (trimetric)	**786.** (metallic)
787. (heavy)	**788.** (non- foliate)
789. (tectosilicate)	**790.** (32 times)
791. (humid)	**792.** (necessary)
793. (removal)	**794.** (red)
795. (Solum)	**796.** (Podzolization)
797. (Argillic Pedoturbation)	**798.** (Intrazonal soils)
799. (arid regions)	**800.** (hyper thermic)
801. (black)	**802.** (Class III)
803. (Silt)	**804.** (Type)
805. (smaller)	**806.** (lighter)
807. (non-capillary)	**808.** (non-liquid)
809. (potential)	**810.** (0.5 bars)
811. (reduces)	**812.** (Diffusion)
813. (inversely)	**814.** (less)
815. (more)	**816.** (slower)
817. (15 %)	**818.** (2::1 non-expending)
819. (2::1 type of clays)	**820.** (pollinations)
821. (Kaolinite)	**822.** (Molybdenum)
823. ($H_2PO_4^-$)	**824.** (montmorillonite)
825. (flocculated)	**826.** (lower)
827. (chloride)	**828.** (may not be applied)
829. (reduces)	**830.** (reduces))
831. (organic)	**832.** (older)
833. (atmosphere)	**834.** (non-hygroscopic)

835. (readily available)
836. (soluble)
837. (upward)
838. (Atmosphere)
839. (enhances)
840. (low)
841. (loss)
842. (Mg)
843. (Foliar application)
844. (Zn)
845. (Mg)
846. (COOH group)
847. (magnesite)
848. (increase)
849. (Mo)
850. (Borax)
851. (thermophiles)
852. (Photoautotroph)
853. (algae)
854. (fungi)
855. (non-symbiotic)
856. (deeper layer)
857. (Plant tissue)
858. (wide)
859. (Humic)
860. (Non- humic)
861. (acidic)
862. (negatively)
863. (Bacteria)
864. (Twenty-one)
865. (Hemicelluloses)
866. (Azotobacter)
867. (Azolla)
868. (not beneficial)
869. (herbicides)
870. (aerobic)
871. (toxic)
872. (volatile)
873. (lead)
874. (roots)
875. (stratosphere)

VI. TRUE (T) OR FALSE (F)

876. The neutral soil aggregates called as peds.
877. Acid soil can reclaim by $CaSO_4.2H_2O$.
878. Alkali soil can claim by lime.
879. Soil profile may define as vertical section of soil.
880. Soil air contains 0.02% CO_2.
881. Muriate of potash contains 60% K_2O.
882. De- nitrification is more in heavy soil.
883. Kaolinite is a 1:: 1 type clay mineral.
884. Quartz is an igneous rock.
885. *Azotobacter* is an anaerobic N_2 - fixation Bacteria.
886. *Clostridium is* an anaerobic N_2 - fixation Bacteria.
887. Neutral condition of soil is most suitable for soil organism.
888. The process of conversion of ammonia to nitrite and then to nitrate is known as nitrification.
889. Grey speck of oat disease caused by Mg deficiency.
890. Zinc, boron, copper elements are micronutrients.
891. Calcium, Mg and S elements are secondary nutrients.
892. The composition of humus contains 50 % C, 35 %O, 5% N and 5% H.
893. Fulvic acid is lowest molecules weight and both acid and alkali soluble.
894. Thc soil organisms are most active at 50 -65°C.
895. Humus is in dynamic condition.
896. Major portion of P in humus is in the form of phytin.
897. When compost prepared by using N-fixing bacteria called super compost.
898. Night soil contains 5.5 % nitrogen.
899. Poultry manure contains 1.45 % N_2.
900. Neem cake is a nitrification inhibitor.
901. Bird guano contains 7-8 % N and 11-14 % P_2O_5.
902. Cotton cake contains 13.5 % N_2.
903. Sunflower cake contains 7.8 % N_2.
904. Binary fertilizer contain two major nutrients e.g. potassium nitrate.
905. $CO(NH_2)_2$ contain 49 % N_2.
906. Calcium Ammonium Nitrate called KISHAN KHAD.
907. Kishan Khad contains 25-28 % N_2.

908. Complete fertilizer having all the three primary major nutrients *viz.* N, P and K.

909. Aqueous ammonia contains 80 % N_2.

910. Rock phosphate contains 20-30 % P_2O_5.

911. Potassium Chloride (KCl) contains 60 % K_2O.

912. Gypsum contains 18.6 % Ca and 29.2 % S.

913. FeS_2 contains 22-24 % sulphur.

914. 50% of India soil is deficient in zinc.

915. One packet of *Rhizobium* culture contains 400 g of Rhizobium.

916. BGA can add up to about 40-45 kg N /ha to rice field.

917. *Azotobacter* uses for rice, cotton and sugarcane.

918. Nitrogen, P_2O_5 and K nutrients are highly mobile in plants.

919. Al, Fe, Mn and Zn nutrients are available at low pH.

920. P and B are available at neutral pH.

921. Tourmaline is the source of boron.

922. Albite, augite olivine quartz are example of secondary minerals.

923. Dolomite is the source of Mg.

924. Gneiss, marble state and sehis are metamorphic rocks.

925. Dolomite, shale limestone are sedimentary rocks.

926. Granite, basalt gabbro is examples of metamorphic rocks.

927. Acid rock contains 20-30 % silica.

928. Earth's crust contains 49.2 % O_2.

929. 'A' horizon also called illuviation horizon.

930. Clay particles size is less than 0.002 mm in diameter.

931. Dominant minerals in salt are quartz and feldspar.

932. The pF range of capillary water is 8.5-9.0.

933. Allelopathy is the process by which one plant infuses the soil with a chemical that affects the growth of other plants.

934. Oxisols are found an about 80 % of the world land.

935. Humic acid is medium in molecular weight and soluble in alkali but insoluble in acid.

936. C:N ratio of FYM is 40:1.

937. Water logged soils increases the availability of Fe and Mg.

938. Diameter of silt particles is less than 0.002 mm.

939. Bulk density of a soil is always higher than particle density.

940. Clay soils grouped as fine textured class.

941. Pore space found in clay soils are more than sandy soils.

942. $pH = -\log \frac{1}{H^+}$

943. The reserve acidity is lower than active acidity

944. The most abundant soil found in India is alluvial soil.

945. *Aspergillus niger* test is used for Cu and Zn.

946. Peat soils have 10-40 % organic matter.

947. The lithosphere mainly made up of igneous rocks.

948. Calcium is not secondary nutrients.

949. Zinc is a micro nutrients

950. Molybdenum is ultra micronutrients.

951. Granite is the sedimentary rocks.

952. $CaCO_3$ add to reclaim of alkali soil.

953. Specific heat of the dry soil is equal to 1.5 al /g.

954. NBSS and LUP are located at Hyderabad.

955. Fertigation is the method of irrigation with fertilizers.

956. 2, 4-D included in phenoxy fatty acids groups.

957. Neutron probe means an instrument used to probe into the plant body using neutrons.

958. The sunflower seeds contain maximum unsaturated fatty acids.

959. Groundnut seed contains about 45% oil and 26% protein.

960. Green manuring refers to manuring when the crop is green.

961. The simazine recommended weedicide to maize.

962. The crop requiring highest water requirement is paddy.

963. *Rhizobium melilotii* culture helps in nitrogen fixation in *Medicago sativa* L.

964. M.H. is growth regulators.

965. Long-term effect of urea application in soils is Acidic.

966. Red soils have the highest cation exchange capacity.

967. The gas-generated form the biogas plant is methane.

968. The maximum available soil water storage in silty clay loam soil.

969. Harmful substance in urea is biuret

970. Microorganism for fixation of atmospheric nitrogen requires potassium

971. Azotobacter brings the conversion of nitrite to nitrates.

972. CAN fertilizers is most suited for foliar spray.

973. Nitrogen deficiency symptoms in plants first appear on lower leaves.

974. In Rice crop, blue green algae are the most useful.

975. Prorate is carbonate insecticide.

976. Darcy's law describes flow of water in saturated soil.

977. Height of rise of water in a capillary tube is directly proportion to radius of capillary tube and viscosity of water.

978. True principle density of most soil is usually 1.35 g / $cm^{3.}$

979. Ca increases nitrogen fixation in soil by Rhizobium.

980. Maximum Available water holding capacity is in silty clay loam.

981. Magnesium in octahedral sheet of clay mineral indicates trioctahedral.

982. Size of colloidal clay particle is less than 0.002 mm.

983. The use of tensiometer confine up to metric potential of – o.8 bar.

984. In the routine soil testing, available nitrogen estimates indirectly by determining total nitrogen.

985. Alkali soil can reclaim by the addition of lime.

986. The available water for plants present in soil is in the between field capacity and wilting point.

987. The arrangement of sand, silt and clay particles in definite set of pattern in soil called texture.

988. Dominant mineral in igneous rock is quartz.

989. Soil aeration units plants growth when percent pores filled with water are more than 50-60 %.

990. Each unit change in pH represents a change in activity of H^+ or OH^- ions by ten folds.

991. Humic acid contains the different elements in order of C > O> H> N.

992. In soil colour determination, chrome refers purity or strength of colour.

993. Gypsum applies in mustard to supply of Iron.

994. Nitrogen fixed in leguminous crops by BGA.

995. "Browning" in cauliflower is due to boron deficiency.

996. Nitrogen fixation through BGA takes place in a paddy crops.

997. Saline alkali soils generally, are in flood regions.

998. The amino acid causes pungency in mustard oil.

999. Phosphorus is a macronutrient due to available in plant in relatively large amount.

1000. PSB improves soil fertility due to releasing foxed phosphorus in the soil.

Answers

876. (T)	**877.** (F)	**878.** (F)	**879.** (T)
880. (F)	**881.** (T)	**882.** (F)	**883.** (T)
884. (T)	**885.** (F)	**886.** (T)	**887.** (T)
888. (T)	**889.** (F)	**890.** (T)	**891.** (T)
892. (T)	**893.** (T)	**894.** (F)	**895.** (T)
896. (T)	**897.** (F)	**898.** (T)	**899.** (F)
900. (T)	**901.** (T)	**902.** (F)	**903.** (T)
904. (T)	**905.** (F)	**906.** (T)	**907.** (T)
908. (T)	**909.** (T)	**910.** (T)	**911.** (T)
912. (F)	**913.** (T)	**914.** (T)	**915.** (F)
916. (F)	**917.** (T)	**918.** (T)	**919.** (T)
920. (T)	**921.** (T)	**922.** (F)	**923.** (T)
924. (T)	**925.** (T)	**926.** (F)	**927.** (F)
928. (T)	**929.** (T)	**930.** (T)	**931.** (T)
932. (F)	**933.** (T)	**934.** (F)	**935.** (T)
936. (T)	**937.** (T)	**938.** (F)	**939.** (F)
940. (T)	**941.** (F)	**942.** (T)	**943.** (T)
944. (T)	**945.** (F)	**946.** (T)	**947.** (T)
948. (F)	**949.** (T)	**950.** (T)	**951.** (F)
952. (F)	**953.** (F)	**954.** (F)	**955.** (T)
956. (T)	**957.** (F)	**958.** (T)	**959.** (T)
960. (F)	**961.** (T)	**962.** (T)	**963.** (T)
964. (T)	**965.** (T)	**966.** (F)	**967.** (T)
968. (F)	**969.** (T)	**970.** (F)	**971.** (F)
972. (T)	**973.** (T)	**974.** (T)	**975.** (T)
976. (T)	**977.** (T)	**978.** (F)	**979.** (F)
980. (T)	**981.** (F)	**982.** (T)	**983.** (T)
984. (T)	**985.** (F)	**986.** (T)	**987.** (F)
988. (T)	**989.** (F)	**990.** (T)	**991.** (T)
992. (F)	**993.** (F)	**994.** (F)	**995.** (T)
996. (T)	**997.** (F)	**998.** (F)	**999.** (F)
1000. (T)			

Bibliography

S. No.	Name of Journal	Country	Jnl Id
1.	ANGRAU Journal of Research	India	AO4
2.	Agriculture Research Journal of Kerala	India	AO15
3.	Agropedology	India	AO67
4.	Analytical Chemistry	India	A100
5.	Annals of Agriculture Research	India	A117
6.	Annals of Plant and Soil Research	India	-
7.	Annals of Arid Zone	India	A121
8.	Applied Clay Science	India	A153
9.	Applied Soil Ecology	India	A164
10.	Archives of Agronomy and Soil Science	Foreign	A182
11.	Arid Soil Research and Rehabilitation	USA	A194
12.	Australian Journal of Agriculture Research	Australia	A210
13.	Australian Journal of Soil Research	Australia	A220
14.	Australian Journal of Soil Science	Australia	A221
15.	Australian Journal of Soil and Water Research	Australia	A222
16.	Bioorganic Chemistry	Foreign	BO76
17.	Biology and Fertility of Soils	USA	
18.	Biological Trace Element Research	UK	
19.	Canadian Journal of Chemistry	Canada	C007
20.	Canadian Journal of Plant Science	Canada	C012
21.	Canadian Journal of Remote Sensing	Canada	C013
22.	Canadian Journal of Soil Science	Canada	C014

S. No.	Name of Journal	Country	Jnl Id
23.	Chinese Agriculture Sciences	China	C043
24.	Clay Minerals	Foreign	C053
25.	Clays and Clays Minerals	Foreign	C054
26.	Critical Review in Plant Sciences	Foreign	C078
27.	Communications in Soil Science and Plant Analysis	USA	
28.	Development in Plant and Soil Science:: Plant Nutrition-	Foreign	D010
29.	Earth Observation and Remote Sensing	Foreign	E002
30.	Emirates Journal of Agriculture Sciences	Foreign	E021
31.	European Journal of Soil Biology	Foreign	E087
32.	European Journal of Soil Science (Journal of Soil Science)	Foreign	E088
33.	Food Chemistry	UK	
34.	Geoderma	Foreign	G022
35.	India Journal of Agricultural Chemistry	India	I019
36.	India Journal of Agricultural Sciences	India	I022
37.	India Journal of Chemistry Section A	India	I033
38.	India Journal of Chemistry Section A	India	I034
39.	India Journal of Earth Science	India	I040
40.	India Journal of Soil Conservation	India	I079
41.	International Journal of Remote Sensing	Foreign	I133
42.	Israel Journal of Plant Sciences	Foreign	I156
43.	JNKVV Research Journal	India	J001
44.	Journal of Agricultural Sciences	Cuba	J014
45.	Journal of Agriculture and food chemistry	USA	-
46.	Journal of the American Oil Chemistry Society	USA	
47.	Journal of Colloid and Interface Science	Foreign	J093
48.	Journal of India Society of Remote Sensing	India	J171
49.	Journal of India Society of Soil Science	India	J172
50.	Journal of Organic Chemistry	India	J213
51.	Journal of Plant Nutrition and Soil Science	India	J252

S. No.	Name of Journal	Country	Jnl Id
52.	Journal of Plant Research	India	J254
53.	Journal of Potassium Research	India	J257
54.	Journal of Soil Conservation	Foreign	J278
55.	Journal of Soil Contamination	Foreign	J279
56.	Journal of Soil Science(Cambridge)	Foreign	J280
57.	Journal of the Science of Food and Agriculture	UK	
58.	International Journal of Food Sciences and nutrition	UK	
59.	Journal of Soil and Water Conservation	India	J281
60.	Journal of Soil and Water Conservation	India	J282
61.	Karnataka Journal of Agricultural Sciences	India	K001
62.	Netherlands Journal of Agricultural Sciences	Netherlands	N018
63.	Orissa Journal of Agricultural Research	India	O006
64.	Plant and Soil	Netherlands	P073
65.	Soil and Plant Science	Foreign	S019
66.	Soil Biology and Biochemistry	UK	S020
67.	Soil Chemistry	Foreign	S021
68.	Soil Science	Foreign	S023
69.	Soil Science Society of America Journal	USA	S024
70.	Soil Science and Plant Nutrition	Foreign	S025
71.	Soil Use and Management	Foreign	S026
72.	Soil and Tillage Research	Netherlands	S027
73.	Water ,Air and Soil pollution	Foreign	W010

Books

S.No.	Name of Books	Authors	Published from
1.	Text Book of Soil Science	T.D. Biswas and S.K. Mukhererjee	Tata McGraw-Hill Publishing Company Ltd,New Delhi
2.	Soil fertility and Fertilizers	Samuel L. Tisdale, Werner L.Nelson, James D. Beaton and John L. Havlin	Prentice-Hall of India Pvt Ltd M-97 Connaught Circus New delhi
3.	Manures and Fertilizers	P.C.Das	Kalyani Publishers,New Delhi
4.	Soils and fertilizers	LL Somani and P.C. DAS	Agrotech publishing Academy Udaipur
5.	Fertilizers-A text book	Ranjan Kumar Basak	Kalyani Publishers,New Delhi
6.	Hand Book of Soil Science	Malcolm E. Sumner	CRC press New York
7.	A text book of Soil Science and Fertilizer Manures	Lal Singh Nirankari	Aman Publishing House Meerut (UP)
8.	Efficient Use of Fertilizers	L.L. Somani	Agrotech publishing Academy Udaipur
9.	Soils in India	P.C Das	Kalyani Publishers,New Delhi
10.	Text book of Soil Science	R.K.Mehra	Directorate of Information and Publication of Agriculture ,New Delhi
11.	Pedology,-(Concept and Application)	J.Sehgal	Kalyani Publishers,New Delhi
12.	Manures and Fertilizers	KS Yawalkar, JP Agarwal and S. Bokde	Agri-Horticultural Publishing House Nagpur
13.	Irrigation and Drainage	D.Lenka	Kalyani Publishers,New Delhi
14.	Agrochemicals and Sustainable Agriculture	NK Roy	APC Publication Pvt Ltd. New Dehi
15.	Analytical Chemistry-theory and practice	R.M. Verma	CBS publishers & Distributors New Delhi
16.	Soil Physics	M.C. Oswal	Oxford & IBH Publishing Co. Pvt .Ltd. New Delhi
17.	Soil Fertility and Nutrient Management	S.S. Singh	Kalyani Publishers,New Delhi
18.	Hand book of Agricultural Sciences	S.S. Singh P.Gupta and A.K. Gupta	Kalyani Publishers,New Delhi

S.No.	Name of Books	Authors	Published from
19.	Introduction to Soil Microbiology	Martin Alexander	John Wiley & Sons Inc.,New York and Landon
20.	Soil Microbiology	N.S. Subba Rao	Oxford & IBH Publishing Co. Pvt .Ltd. New Delhi
21.	The Nature and Properties of Soil Science	N.C. Brady	–
22.	A Hand Book of Organic Farming	A.K Sharma	Agro-bios , Jodhpur (India)
23.	Soil, Plant ,Water and Fertilizer Analysis	P.K Gupta	Agro-bios , Jodhpur (India)
24.	Methods of Analysis of soil, Plants ,Water and Fertilizers	HLS Tandon	FDCO, New Delhi
25.	Handbook of Agriculture	-	Directorate of Information and Publication of Agriculture ,new Delhi
26.	Fundamentals of Soil Science	-	India Society of Soil Science
27.	Soil less growth of Plants	C. Illis	J.V. Publishing House ,Jodhpur
28.	Soil Fertility and Sewage	JPJ, Van Vuren	J.V. Publishin House, Jodhpur
29.	Soil and Soil Management	A.F. Gustafson	J.V. Publishing House, Jodhpur
30.	A Hand book of Mineral Nutrition and Diagnostic Techniques for nutritional Disorders of Crops	G. Pathmanabhan	Agro-bios , Jodhpur (India)
31.	Analysis of Soil Physics properties	S.P. Majumdar	Agro-bios , Jodhpur (India)
32.	A handbook of Soil ,Fertilizer and Manure	P.K. Gupta	J.V. Publishing House, Jodhpur
33.	Crop Production with Saline Water	L.L. Somani	J.V. Publishing House, Jodhpur
34.	Synthetic Fertilizers for Sustainable Agriculture	R.A.Raju	Agro-bios , Jodhpur (India)
35.	Soil Conservation and Fertility Management	S.C. Panda	Agro-bios , Jodhpur (India)

Glossary

A horizon: The surface Horizon of a mineral soil having maximum organic matter accumulation, maximum biological activity, and / or eluviation of materials such as iron and aluminium oxides and silicates clays.

ABC soil: A soil with a distinctly developed profile including A, B and C-horizons

Available Water: The portion of water in a soil that can be readily absorbed by plant roots The amount of water released between the field capacity and the permanent wilting point.

Acid soil: A soil that is acid in reaction throughout the root zone, a soil with has more hydrogen ion than hydroxyl ions in the soil solution, especially soil with pH value less than 6.6. Acid soils are reclaimed by addition of liming materials.

Additive: A material added to fertilizer to improve its physical condition.

Alkali soil: A soil that is alkaline in reaction throughout the root zone, any soil having a pH value greater than 7.0. Practically, a soil having pH above 7.3 is called alkaline soil. An alkaline soil can be reclaimed by the use of gypsum or sulphur.

Autotroph: An organism capable of utilizing carbon dioxide or carbonates as the sole source of carbon and obtaining energy for life process from the oxidation of inorganic elements like Fe, Al, S H_2 and nitrate.

Aggregates: many soil particles held in a single mass or cluster such as clod, crumb, block, or prism.

Alkalization: The process in which the exchangeable sodium content of a soil is increased.

Alluvial soil: A soil developing from recently deposited alluvium and exhibiting essentially no horizon development or modification of the recently deposited materials.

Amendment, soil: Any materials such as lime, gypsum saw dust is added to the soil to change the properties of a soil.

Alumino-silicates: Compounds containing Al, Si, and O_2 as main constituents.

Ammonification: The biochemical process whereby ammoniacal nitrogen is released from nitrogen containing organic compounds.

Ammonium fixation: The absorption of ammonium ions by the mineral or organic fractions of the soil.

Antibiotic: A substance produced by one species of organism that in low concentration will kill or inhibits growth of certain other organism.

Azonal soil: Soils without distinct genetic horizon, a soil order under the 1949 classification system.

B-horizon: A soil horizon usually beneath the A, which is characterized by one or both of the following: (1) an accumulation of silicates clays iron and aluminium oxides and humus alone or in combination; (2) a blocky or prismatic structure.

Bar A unit of pressure equal to one million dynes per square centimeter.

Band placement: The placement of fertilizers in the soil in continuous band or narrow ribbon in between the plant rows or around each plant. Phosphatic fertilizers are normally placed in bands to reduce phosphoric fixation.

Bedding soil: Arranging the surface of fields by plowing and grading into a series of elevated beds separated by shallow depression or ditches for drainage.

Bedrock: The solid rock underlying soils and the Regolith in depths ranging from zero to several hundred feet.

Bulk density of soil: The mass of dry soil per unit of bulk volume including the air space.

Buffering capacity: The ability of a soil to resist change in pH commonly determined by presence of clay, humus and other colloidal materials.

Broadcast: Scatter seed or fertilizer on the surface of soil.

Bio-solids: Sewage sludge that meets certain regulatory standards making it suitable for land application.

Bio sequence: A group of related soils that differ one from the other primarily because of differences in kinds and numbers of plants and soil organisms as a soil-forming factor.

Bioremediation: The decontamination or restoration of polluted or degraded soils by means of enhancing the chemical degradation or other activities of soil organism.

Basic slag: A by-product of steel industry obtained from phosphate iron ores. It contains about 6-18 % total phosphoric acid and 5 to 15 % in available form. Basic slag produced in India being of low grade is not used as a commercial fertilizer.

Bio pores: Soil pores usually of relative large diameter created by plant roots earthworms or other soil organism.

C-horizons: A horizon generally beneath the solum, which is relatively little affected by biological activity.

Concretion: A local concentration of a chemical compound such as calcium carbonate or iron oxide in the form of grain or nodules of varying size shape hardness and colour.

Conduction: The transfer of heat by physical contact between two or more object.

Consistency: The interaction of adhesive and cohesive forces within a soil at various moisture contents as expressed by the relative case with which the soil can be deformed.

Consistence: The combination of properties of soil material that determine its resistance to crushing and its ability to be molded or change in shape such terms as loose friable firm soft plastic and sticky describe soil consistence.

Consumptive Use: The water used by plants in transpiration and growth plus water vapour loss from adjust soil or snow or from intercepted.

Carbon: Nitrogen Ratio: The ratio of the weight of organic carbon (C) to the weight of total nitrogen (N) in a soil or in organic material.

Catena: A sequence of soils of about the same age, derived from similar parent material and occurring under same similar climate conditions, but having different characteristics due to variation in relief and in drainage.

Chelate: A typc of chemical compound in which a metallic ion is firmly combined with a molecule by means of multiple chemical bonds.

Cryoturbation: Physical disruption and displacement of soil materials within the profile by the forces of freezing and thawing sometimes called frost churning.

Clay: Soil material containing more than 40% clay, less than 45 % sand and less than 40 % silt.

Clayey: Containing large amounts of clay or having properties similar to those of clay.

Clay mineral: Naturally occurring inorganic material usually crystalline found in soils, the particles being of clay size; that is < 0.02 mm in diameter.

Clay pan: A compost slowly permeable layer in the sub soil having a much higher clay content than the overlaying materials from which it is separated by a sharply defined boundary. Clay pan are usually hard when dry and plastic and sticky when wet.

Desert Pavement: A natural residual concentration of closely packed pebbles boulders and other rock fragments on desert regions.

Diatomaceous Earth: A geologic deposit of fine grayish siliceous material composed of chiefly or wholly of the remains of diatoms.

Double layer: In colloid chemistry, the electric charges of the surface of the disperse phase and the adjacent diffuse layer of ions in solution.

Drift: Material of any sort deposited by geological processes in one place after having been removed from another. Glacial drift includes materials moved by the glaciers.

Drumlin: Long, smooth cigar shaped low hills of glacial till with their long axes parallel to the direction of ice movement.

Dust mulch: A loose finely granular or powdery condition on the surface of the soil usually produced by shallow cultivation.

Duripan: A diagnostic sub-horizon that is cemented by silica to the point that air-dries fragments will not slake in water or HCl hardpan.

E-horizon: Horizon characterized by maximum illuviation or 'washing out' of silicate clays, iron, and aluminum oxides commonly occurs above the B-horizon and below the A-horizon.

Eluviation: The removal of soil material in suspension or in solution from a layer or layers of a soil.

Equivalent acidity: A term used to express the calcium carbonate equivalent of acidic residue left by a fertilizer material.

Equivalent basicity: A term used to express the calcium carbonate of basic residue left by a fertilizer material.

Equilibrium Phosphorus Concentration: The concentration of P in a solution in equilibrium with a soil the EPCo being the concentration of P achieved by desorption of P from a soil to P free distilled water.

Eutrophic: Having concentrations of nutrients optimal for plant or animal growth

Exfoliation: Peeling away of layer of a rock from the surface inward usually as the result of expansion and contraction that company changes in temperature.

Expensive soil: Soil that undergoes significant volume change upon wetting and drying usually because of a high content of swelling type clay minerals.

Fertigation: The application of fertilizers in irrigation waters commonly through sprinkler systems.

Flagstone: A relative thin rock or mineral fragment 15-38 cm in length commonly composed of shale slate limestone or sandstone.

Friable: A soil consistency term pertaining to the ease of crumbling of soils.

Frigid: A soil temperature class with mean annual temperature below 8^0C.

Gilgai: The micro relief of soil produced by expansion and contraction with change in moisture.

Genetic Horizon : Soil layers that resulted from soil forming as apposed to sedimentation or other geologic processes.

Green leaf manuring: This refers to turning under of green leaves and tender green twigs collected from shrubs and trees grown on bunds wasteland s and nearly forest areas.

Gypsum Requirement: The quantity of gypsum required to reduce the exchangeable sodium percentage in a soil to a an acceptable level.

Hardpan: A soil with a distinctly developed profile including A, B and C horizons.

Half- bog soil: A great soil group of the Intrazonal order and Hydromorphic suborder consisting of soil with dark brown or black peaty material over grayish and rust mottled mineral soil.

Heavy soil: A soil with a high content of the fine separates particularly clay or one with a high drawbar pull hence, difficult to cultivate.

Horn and hoof meal: Manure prepared from horns and hoofs of animal such a meal contains about 13 % nitrogen.

Heavy soil: A soil with high content of the fine separates, particularly clay or one with a high draw bar pull and hence, difficult to cultivate.

Hydromorphic soils: A suborder of Intrazonal soils all formed under conditions of poor drainage in marshes, swamps seepage areas, or flats (1949 classification system).

Immature soil: A soil with indistinct or only slightly developed horizons because of the relatively short time it has been subjected to the various soil forming processes.

Indurate soil : Soil material cemented into the hard mass that will not soften on wetting.

Integrade: A soil that possesses moderately well developed distinguishing characteristics of two or more genetically related great soil groups.

Ions: Electrically charged particles; positively charged ions are called 'cation' and negatively charged particles are called 'anions'

Kame: An irregular ridge or hill of stratified glacial drift.

Light soil : A term used for sandy or coarse texture soil. Green manuring should be done to improve physical condition of a light soil.

Lithosols: a great soil group of azonal soils characterized by an incomplete solum or no clearly expressed soil morphology and consisting of freshly and imperfectly weathered rock or rock fragments.

Loam: The soil having moderate amount of sand, silt, and clay generally loam soil contains 7 to 27 % of clay 28 to 50% of silt and less than 52% of sand.

Loamy: Intermediate in texture properties between fine textured and coarse textured soils.

Loess: Materials transported and deposited by wind and consisting of predominantly silt-sized particles.

Luxury Consumption: The intake by a plant of an essential nutrient in amounts exceeding what is needs. Thus, if potassium is abundant in the soil, alfalfa may take in more than is required.

Lysimeter: A device for measuring percolation and leaching losses from a column of soil under controlled conditions.

Marl: Soft and unconsolidated calcium carbonate, usually mixed with varying amounts of clays or other impurities.

Marsh: Periodically wet or continually flooded areas with the surface not deeply submerged.

Mature soil: A soil with well-developed soil horizons produced by the natural processes of soil formation and essentially in equilibrium with its present environment.

Mellow soil: A very soft, very friable, porous soil without any tendency toward hardness or harshness.

Mineral soil : A soil consisting predominantly of and having its properties determined predominantly by mineral matter, usually contains > 20 % OM but may contain an organic surface layer up to 30 cm.

Mor: Raw humus ; a type of forest humus layer of unincorporated organic material usually matted or compacted or both distinct from the mineral soil unless the latter has been blackened by washing in organic matter.

Muck soil: A soil containing 20-50% organic matter or an organic soil in which the organic matter is well decomposed.

Mull: A humus rich layer of forested soils consisting of mixed organic and mineral matter.

Neutral soil: A soil in which the surface layer at least to normal plow depth is neither acid nor alkaline in reaction in practice this means the soil is within the pH range of 6.6 to 7.3.

Nitrate Depression Period: A period of time beginning shortly after the addition of fresh highly carbonaceous organic materials to a soil during which decomposer microorganisms have removed most of the soluble nitrate from the soil solution.

Night Soil: Night soil is human excreta, solid and liquid. On average, Night soil contains 5.5% Nitrogen, 4.0 Phosphorus (P_2O_5) and 2.0 % Potash (K_2O) on oven dry basis.

O-horizon: Organic horizon of mineral soil.

Organic soil: A soil which contains a high percentage (> 20%) of organic matter thought out the solum.

Ortstein: An indurate layer in the B-horizon of Podzol / spodosols in which the cementing material consists of illuviated sesquioxide.

Oxisols: Orders in soil taxonomy soils with residual accumulations of low activity clays free oxides kaolin and quartz they are mostly in tropical climates.

Peat: Unconsolidated soil material consisting largely of undecomposed or slightly decomposed, organic matter accumulated under conditions of excessive moisture.

Ped: A unit of soil structure such as an aggregate crumb, prism, block or granule formed by natural process(in contrast with a clod, which is formed artificially).

Pedalfer: A subdivision of a soil order comprising a large group of soils in which sesquioxides increased relative to silica during soil formations.

Pedocals: A subdivision of a soil order comprising a large group of soils in which calcium accumulated during soil formation.

Permafrost: Permanently frozen materials underlying the solum.

pF: The logarithm of the soil moisture tension expressed in cm height of a column of water.

pH: The negative logarithm of the hydrogen ion activity of a soil.

Photomap: A mosaic map made from aerial photographs with physical and cultural features shown as on a planimatric map.

Planosol: A great soil group of the Intrazonal order and Hydromorphic suborder consisting of soil with Eluviated surface horizons underlain by B-horizons more strongly eluviated cemented or compacted than associated normal soil.

Plastic soil: A soil capable of being molded or deformed continuously and permanently, by relatively moderately pressure, into various shapes.

Prairie soils: The volume percentage of the total soil bulk not occupied by solid particles.

Porosity soil: A soil with a distinctly developed profile including A,B and C horizons.

Peddled soil: Dense massive soil artificially compacted when wet and having no aggregated structures, the condition commonly results from the tillage of a clayey soil when it is wet.

Quick Soil Test: These are simple and rapid chemical test of soils designated to give an approximation of the nutrients available to plants.

Regolith: The unconsolidated mantle of weathered rock and soil material on the earth surface loose earth materials above soil rock equivalent to the term soil as use many engineers.

Regosol: Any soil of the Azonal order without definite genetic horizons and developing from or on deep, unconsolidated soft mineral deposits such as sands, loess, glacial drift.

Rill: A small intermittent water course with steep sides; usually only a few inches deep and hence, no obstacle to tillage operations.

Saltation: Particles movement in water or wind where particles skip or bounce along the streambed or soil surface.

Self Mulching Soil: A soil in which the surface layer becomes so well aggregated that it does not crust and seal under the impact of rain but instead serves as surface mulch upon drying.

Separate soil: One of the individual sized groups of mineral soil particles sand silt and clay.

Series soil: The soil series is a subdivision of a family in soil taxonomy and consist of soils that are similar in all major profile characteristics.

Solonetz soils: An Intrazonal group of soils having surface horizons of varying degree of friability underlain by dark hard soil, ordinarily with columnar structure (prismatic with rounded tops). This hard layer is usually highly alkaline. Such Soils are developed under grass or shrub vegetation mostly in sub humid and semiarid climates.

Soil Arc : The soil atmosphere the gaseous phase of the soil being that volume not occupied by soil or liquid.

Stratified: Arrange in or composed of strata or layers.

Talus: Fragmented of rocks and other soil materials accumulated by gravity at the foot of cliffs or steep slops.

Toposaic: A photomap on which topographic or terrain form lines are shown as on standard topographic quadrangles.

Topsequence: A sequence of related soils that differ one from other primarily because of topography as a soil formation factor.

Truncated : Having lost all or part of the upper soil horizon or horizons.

Tuff: Volcanic ash usually more or less stratified and in various states consolidated.

Tundra soil: A zonal great soil group consisting of soils with dark brown peaty layers over grayish horizons mottled with rust and having continually frozen substrata ; formed under frigid humid climate with poor drainage and native vegetation of lichens moss flowering plants and shrubs.

Thixotrophy: The properties of certain clay soil of becoming fluid when formed or agitated and then setting again when at rest similar to quick as in quick clays or quicksand.

Virgin soil: A soil that has not been significantly disturbed from its natural environment.

Water Use Efficiency: Dry matter or harvested portion of crop produced per unit of water consumed.

Weathering: All physical and chemical changes produced in rocks at or near the earth's surface by atmospheric agents.

Xerophytes: Plants that grow in or on extremely dry soils or soil materials.

Zonal soil: A soil characteristics of a large area or zone.

Appendices

Appendix-1

Molecular and equivalent weights of some important compounds

Compound	Formula	Mol. Wt (g)	Eq.Wt(g)
Ammonium acetate	CH_3COONH_4	77.08	77.08
Ammonium chloride	NH_4NCl	53.49	53.49
Ammonium fluoride	NH_4F	37.04	37.04
Ammonium nitrate	NH_4NO_3	80.04	80.04
Acetic acid	CH_3COOH	60.05	60.05
Barium acetate	$(CH_3COO)_2Ba$	255.43	127.72
Barium chloride dihydrate	$BaCl_2.2H_2O$	244.28	122.14
Boric acid	H_3BO_4	61.83	20.61
Calcium acetate	$(CH_3COO)_2Ca$	158.00	79.00
Calcium carbonate	$CaCO_3$	100.09	50.05
Calcium chloride	$CaCl_2.2H_2O$	147.02	73.51
Calcium hydroxide	$Ca\ (OH)_2$	74.00	37.00
Calcium nitrate	$Ca\ (NO_3)_2$	164.00	82.00
Calcium sulphate	$Ca\ SO_4.2H_2O$	172.17	86.08
Copper sulphate pent hydrate	$CuSO_4.5H_2O$	172.17	86.08
EDTA	$\{CH_2N\ (CH_2COOH)_2\}_2$	295.25	73.06
Ferrous ammonium sulphate	$(NH_4)2SO_4.FeSO_4.6H_2O$	392.13	392.13
Ferrous sulphate	$FeSO_4.7H_2O$	278.01	139.00
Hydrochloric acid	HCl	36.46	36.46
Magnesium chloride	$MgCl_2.6H_2O$	203.30	101.65
Magnesium nitrate	$Mg\ (NO_3)_2.6H_2O$	256.41	128.20
Nitric acid	HNO_3	63.01	63.01
Oxalic acid	$C_2H_2O_4.2H_2O$	126.00	63.00
Potassium chloride	KCl	74.55	74.55
Potassium dichromate	$K_2Cr_2O_7$	294.19	49.04
Potassium hydroxide	KOH	56.10	56.10
Potassium permanganate	$KMnO_4$	158.03	31.60
Potassium nitrate	KNO_3	101.10	101.10

Contd...

Compound	Formula	Mol. Wt (g)	Eq.Wt(g)
Potassium sulphate	K_2SO_4	174.27	87.13
Potassium hydrogen phthalate	COOH C_6H_4COOK	204.22	204.22
Silver nitrate	$AgNO_3$	169.87	169.87
Sodium acetate	CH_3COONa	82.04	82.04
Sodium bicarbonate	$NaHCO_3$	84.01	84.01
Sodium carbonate	Na_2CO_3	106.00	53.00
Sodium chloride	NaCl	58.45	58.45
Sodium hydroxide	NaOH	40.00	40.00
Sodium nitrate	$NaNO_3$	84.99	84.99
Sodium oxalate	$Na_2C_2O_4$	134.00	67.00
Sodium sulphate	Na_2SO_4	142.04	71.02
Sodium thiosulphate	$Na_2S_2O_3.5H_2O$	248.18	248.18
Sulphuric acid	H_2SO_4	98.08	49.04

_Data source: Soil plant water analysis: Dhyan Singh *et. al.*

Appendix-II

Conversion table

1 inch	=	2.54 cm	1 acre inch	=	22,660 gallon
1 foot	=	0.3048 m	1 ha cm	=	100,000
1 m	=	3.28 ft	L	=	10.27 ha cm
1 mile	=	1.60931 km	1 cusec	=	6.25 gallon / sec
1 km	=	0.62 miles		=	102.7 m
1 $inch^2$	=	7.54 cm^2		=	22,500 gallon / hr
1 ft^2	=	9.29-03cm^2	NO_3	=	N X 4.43
1 $mile^2$	=	640acres = 2.5899km^2		=	3600 ft^3 / hr
1 km^2	=	100 ha	OM	=	OC X 1.724
1 ha	=	10,000 m^2= 2.4710 acre	NH_4	=	N X 1.286
1 cubic inch (in^3)	=	16.38 cm^3	P_2O_5	=	P x 2.29
1 ft^3	=	0.02831 m^3	1 cusec / hr	=	1 acre inch
1 lb (pound)	=	453.59 g	K2O	=	K x 1.20
1 ton (t)	=	1000 kg	1 cusec/ day	=	1.98 acre ft
1 quintal	=	100 kg	CaO	=	Ca x 1.40
1 gallon	=	10 lb (pound)	1 atmosphere	=	1.013 x 10^6 dynes
1 ft^3	=	28.31 litre	Ca $(OH)_2$	=	Ca x 1.85
1 m^3 water	=	1000 liter	1 mm sieve	=	16 mesh /inch
	=	35.32 ft^3	SO_4	=	S X 3
	=	219.97 gallon	ppm	=	µg mL^{-1}
1 gallon	=	4.54596 litre	ppm	=	equivalent x meL^{-1}
1 acre inch	=	3630 ft^3	kg /hah	=	1.121x lb $acre^{-1}$
1 ha cm	=	100 m^3	ppm	=	mg / kg
			1 inch	=	2.54 cm

_Data Source: Manures and Fertilizers: PC Das

Appendix-III

Rating chart for soil test values of essential nutrients

Nutrients	Compounds (Units)	Rating		
		Low	Medium	Highs
C:	Organic carbon (%)	< 0.4	0.4 - 0.75	> 0.75
N:	Alkaline $KMnO_4$ (kg/ha)	< 280	281-560	> 560
P:	Olsen's	(Kg/ ha)	< 12.5	12.5-25.0
> 25.0				
K:	Ammonium acetate (kg/ha)	< 135	135-335	> 335
Ca:	Ammonium acetate (cmol (P^{+}) / kg)	< 1.5	———	> 1.5
Mg:	Ammonium acetate (cmol (P^{+}) / kg)	< 1.0	———	> 1.0
Zn:	DTPA extractable (mg/kg)	< 0.6	0.6-1.20	> 1.2
Fe:	DTPA extractable (mg/kg)	< 4.5	4.5-9.0	> 9.0
Mn:	DTPA extractable (mg/kg)	< 3.5	3.5-7.0	> 7.0
Cu:	DTPA extractable (mg/kg)	< 0.2	0.2 – 0.4	> 0.4
B:	Hot water soluble (mg/kg)	< 0.5	0.5 – 1.0	> 1.0
Mo:	Ammonium oxalate extractable (mg/kg)	< 0.2	0.2 - 0.4	> 0.4
S:	0.01M $CaCl_2$ (kg/ha)	< 22.4	22.4-35.0	> 35.0

Source: Hand Book of Agriculture: ICAR

Conductivity: Total Soluble Salt (in milliohms)

Below 1	:	Normal
1- 2	:	Critical for germination
2 – 3	:	Critical for the growth of salt Sensitive
Above 3	:	Injurious to most crops

Appendix-IV

Amount of nutrient in Manures / fertilizers (in percentage)

Name of Manures	Nitrogen (N)	Phosphorus(P_2O_5)	Potassium (K_2O)
1. Bulky Organic Manure			
Farm yard Manure	0.5-1.5	0.4-0.8	0.5-1.9
Compost (urban)	1.0-2.0	1.0	1.5
Compost (Rural)	0.4-.08	0.3-0.6	0.7-1.0
2. Green Manure			
Dhaincha	0.62	0.15	0.58
Sunhemp	0.75	0.12	0.15
Cowpea	0.71	0.15	0.58
3. Oil Cakes			
Mustard cake	5.1-5.2	1.8-1.9	1.1-1.2
Groundnut cake	7.0-7.3	1.5-1.6	1.3-1.4
Sesame or till cake	6.2-6.3	2.0-2.1	1.2-1.3
Linseed cake	5.5-5.6	1.4-1.5	1.2-1.3
Castor cake	4.3	1.8	1.3
Neem cake	5.2-5.3	1.0-1.1	1.0-1.5
Coconut cake	3.0-3.2	1.9-2.0	1.7-1.8
Mahua cake	2.5-2.6	0.8-0.9	1.8
4. Bulky Organic Manure			
Farm Yard Manure	0.5-1.5	0.4-0.8	0.5-1.9
Compost (urban)	1.0-2.0	1.0	1.5
Compost (Rural)	0.4-.08	0.3-0.6	0.7-1.0
5. Manures of Animal Origin			
Bone meal (Raw)	3.0-4.0	20.0-25.0	-
Bone meal (Steamed)	1.0-2.0	25.0-30.0	-
Fish meal	4.0-10.0	3.0-9.0	0.3-1.5
Dried blood	10.0	1.5	1.0
Cow dung (Fresh)	0.3-0.4	0.1-0.2	0.1-0.3
Poultry manure (Fresh)	1.0-1.8	1.4-1.8	0.8-0.9

Name of Fertilizers	Nitrogen (N)	Phosphorus(P_2O_5)	Potassium (K_2O)	
1. Nitrogenous fertilizers				
Ammonium sulphate (AS)	20.6	-	-	
Ammonium Chloride	25.0-26.0	-	-	
Urea	44.0-46.0	-	-	
Calcium ammonium nitrate (CAN)		25.0-26.0	-	-
Ammonium sulphate nitrate (ASN)		26.0	-	-
Calcium nitrate	15.6-25.0	-	-	
2. Phosphatic fertilizers				
Super phosphate (Single)	-	16.0-20.0	-	
Super phosphate (Double)	-	30.0-35.0	-	
Super phosphate (Triple)	-	45.0-50.0	-	
Basic slag	-	3.0-8.0	-	
Rock phosphate (Mussori)	-	23.0-24.0	-	
Rock phosphate (Purulia)	-	23.0	-	
3. Potassium fertilizers				
Muriate of potash	-	-	50.0-60.0	
Potassium sulphate	-	-	48.0-52.0	
Potassium phosphate	-	-	30.0-50.0	
4. Complex fertilizers				
Di-ammonium phosphate	18.0	46.0	-	
Ammonium nitrate phosphate	20.0	20.0	-	
Urea ammonium phosphate	28.0	28.0	-	
Suphala 15:15:15	15.0	15.0	15.0	
Suphala 20:20:0	20.0	20.0	-	
IFFCO 10:26:26Paras 10:26:26	10.0	26.0	26.0	

Source: Hand Book of Agriculture: ICAR

Appendix-V

Fertilizers and its relation to plant nutrients

1 kg Nitrogen	=	5 kg Ammonium Sulphate
	=	4 kg CAN or Calcium Ammonium Nitrate
	=	4 kg Ammonium Chloride
	=	2.22 kg Urea 4 kg ASN or ammonium Sulphate Nitrate
1 kg Phosphate	=	6.25 kg Super phosphate (Single)
1 kg Potash	=	1.660 kg Muriate of Potash (60%) or
	=	2 kg Muriate of Potash (50%)
1 kg Nitrogen+ 1		kg Phosphate = 5 kg Suphala 20:20:20
1 kg Nitrogen +		1 kg phosphate = 3.5 kg Urea Ammonium Phosphate (Gromor 28:28:28)
1 kg Nitrogen +		1 kg phosphate = 1 kg Potash =6,660 kg Suphala 15:15:15

Source: Hand Book of Agriculture: ICAR

Appendix-VI

Material containing Sulphur, Calcium, and Magnesium

Nutrients	Materials	Contents (%)
Sulphur	Single super phosphate	12
	Ammonium sulphate	24
	Potassium sulphate	18
	Ammonium phosphate sulphate	15
	Gypsum	13-18
	Iron pyrites	22-24
Calcium	Agricultural lime stone	80-95 Ca CO_3
	Dolomite	20-45 Ca CO_3
	Gypsum	40 CaO
	Single super phosphate	25-30 CaO
	Rock phosphate	39-48 CaO
	Calcium ammonium nitrate	10-20 CaO

Source: Hand Book of Agriculture: ICAR

Appendix-VII

General Recommended doses of micronutrient fertilizers

Micronutrients	Material and dose for application	
	Soil application	foliar spray
Zinc	Zinc sulphate (10-50 kg /ha)	0.5% zinc sulphate + 0.25 % lime
Iron	Ferrous sulphate (10 kg/ha)	0.4% ferrous sulphate + 0.2 %lime
Copper	Copper sulphate (10-50 kg /ha)	0.1% copper sulphate + 0.05 %lime
Manganese	Manganese sulphate (10-50 kg/ha)	0.4% ferrous sulphate + 0.2 % lime
Zinc	Zinc sulphate (10-50 kg /ha)	0.6% manganese sulphate + 0.25 % lime
Boron	Borax (10-20 kg/ha)	0.2% borax

Data source: Soil plant water analysis: Dhyan Singh et al.

Appendix-VIII
Some indicator plants of nutrients deficiency

Nutrients	Indicator Plants
N	: Maize, cereal (small grain), mustard, apple, citrus
P	: Maize, Barley, lettuce, tomato
K	: Potato, Lucerne, beans, tobacco, cucurbits, cotton, tomato, maize
Ca	: Lucerne, other legume crops
Mg	: Potato, cauliflower
S	: Lucerne, raya
Fe	: Sorghum, Barley, citrus, peach
Zn	: Maize, onion, mustard, peach, citrus
Mn	: Apple, cherry, maize, oats, pea, radish, wheat, tomato
Cu	: Maize, Apple, citrus, barley, lettuce, oats, onion, tobacco, tomato
B	: Lucerne, turnip, cauliflower, apple, peach
Cl	: Lucerne
Mo	: Cauliflower, other *Brasica* spp., citrus, legumes, oats, spinach

Data source: Soil plant water analysis: Dhyan Singh *et. al.*

Appendix-IX
Water quality classification according to USSL

Salinity hazard			Sodium hazard		
Salinity (EC)	**Class**	**EC x 10 ^{3}**	**Sodium hazard**	**Class**	**SAR**
Low	C1	0.01 to 0.25	Low	S1	<10
Medium	C2	0.25 to 0.75	Medium	S2	10 to 18
High	C4	0.75 to 2.25	High	S3	18 to 26
Very high	C5	2.25 to 5.00	Very high	S4	26 to 31
A soil is saline if ECe		> 4 dS/m, ESP < 15 % and pH < 8.5			
A soil is alkali if EC_e		< 4 dS/m, ESP > 15 % and pH > 8.5			
A soil is saline-alkali if EC_e		> 4 dS/m, ESP > 15 % and pH = 8.5(or slightly higher)			

Source: Hand Book of Agriculture: ICAR

Appendix-X
Grouping of poor quality waters

		Water quality class		
Main	**Sub class (dS/m)**	**ECiw (mmo/l)**	**SARiw (meq/l)**	**RSC**
Good	-	<2	<10	<2.5
Saline	Marginally saline	2-4	<10	<2.5
	Saline	>4	<10	<2.5
	High SAR saline	>4	>10	<2.5
Alkali	Marginally alkali	<4	<10	2.5-4.0
	Alkali	<4	<10	>4.00
	High SAR alkali	variable	>10	>4.0

Total cations (or anions), meq/ l = 10 *EC dS/m;

Total salt concentration, mg/l= 640*EC dS/m

Osmotic pressure atmosphere at 25oC = 0.36*EC dS/m; 1dS/m= 1mmhos/cm

Appendix-XI
Irrigation requirement of some important crops

Crops	Crop duration (days)	Irrigation (mm)
Kharif		
Rice	130-140	700-800
Sorghum	110	150
Maize	110	150
Sugarcane	330	700
Summer pulses	70	210
Pearl millet	90	150
Sunflower	110	210
Cotton	180	280
Rabi		
Wheat	135	350
Berseem	200	800
Potato	110	450
Lentil	135	150
Mustard	150	150
Barley	125	210
Soybean	90	350
Chickpea	150	150

Source: Hand Book of Agriculture: ICAR

Appendix-XII
Classification of herbicides based on their chemistry

Group	Examples
Chlorophenoxy compound	: 2, 4-D, MCPA, MCPB, 2, 4-DB
Substituted aliphatic acid	: TCA
Benzoic acids	: Dicamba, Amiben
Anilides	: Propanil
Dinitroanilines	: Trifuralin, Fluchloralin, Pendimethalin, Oryzalin
Substituted ureas	: Diuron, linuron, Isoproturon
Carbamates	: Propham, Chloropepham
Thiocarbamates	: Molinate, triallate EPTC
Triazines	: Atrazine, Prometryn, Metrubuzin, Terbutryn
Uracils	: Terbacil, Bromacil
Organic, arsenicals	: MSMA, Cacodylic acid
Nitriles	: Dichlobenil, Ioxynil, Bromoxynil
Bipyridiliums	: Diquat, Paraquat
Sulphonyl ureas	: Sulphosulfuron, Chlorimuron, ethyl, Metasulfuron- methyl
Phenoxy-phenoxy alkanoic acids	: Clodinafop, fenoxaprop-p-ethyl, Quizalofop
Chloroacetamides	: Alachlor, Butachlor, Metolachlor
Diphenylethers	: Nitrofen, Acifluorfen, Oxyfluorfen
Organophosphorus	: Glyphosate, Anilofos

Source: Hand Book of Agriculture: ICAR

Appendix-XIII
Acid equivalent of acid forming fertilizer

Fertilizers	Acid equivalent
Ammonium Chloride	128
Ammonium Phosphate	86
Ammonium Sulphate	110
Ammonium Sulphate Nitrate	93
Ammonium Nitrate	60
Urea	80
Equivalent basicity of basic fertilizer	
Fertilizers	Equivalent basicity
Calcium Cyanide	63
Calcium nitrate	27
Dicalcium Phosphate	25
Sodium nitrate	29

Data Source: Manures and Fertilizers: PC Das

Appendix-XIV
Optimum pH range for availability of different nutrients

Nutrients	Optimum pH range
N	6.0-8.0
P	6.5-7.5
K	6.0-7.5
Ca	7.0-8.5
Mg	7.0-8.5
Fe	6.0 and below
Mn	5.0-6.5
Bo	5.0-7.0
Cu	5.0-7.0
Zn	5.0-7.0
S	6.0 and above
Mo	7.0 and above

Data Source: Manures and Fertilizers: PC Das